TRAMROADS
OF THE
BRECKNOCK AND ABERGAVENNY CANAL

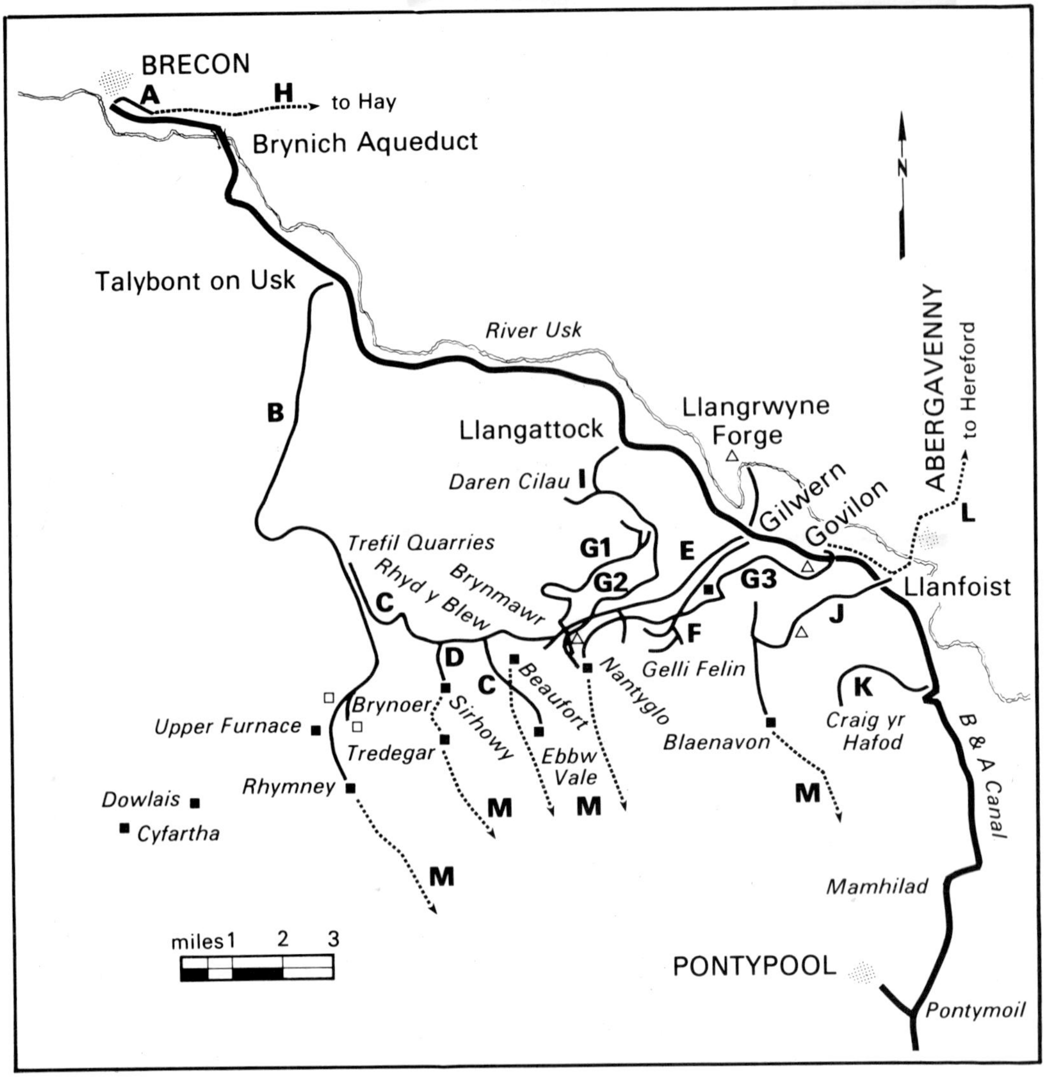

Tramroads of the Brecknock and Abergavenny Canal

A The Watton Plateway
B Brinore Tramroad
C Trevil Railroad
D Rassa Railroad
E Clydach Rail Road
F Llam-march Tramroad
G1 Bailey's Tramroad (Daren Disgwylfa)
G2 Bailey's Tramroad (Llangattock)
G3 Bailey's Tramroad (Govilon)
H Hay Railway
I Llangattock Tramroad
J Hill's Tramroad
K Craig yr Hafod Tramroad
L Llanvihangel Railway

M Monmouthshire Canal Company's Tramroads to Newport

TRAMROADS of the BRECKNOCK & ABERGAVENNY CANAL

GORDON RATTENBURY

RAILWAY AND CANAL
HISTORICAL SOCIETY
1980

with the support of the
South-East Wales Arts Association

First published 1980 by the Railway and Canal Historical Society.
Registered Office: 12 High Street, Oakham, Leicestershire LE15 6AW

ISBN 0 901461 23 7 (Hardback)
ISBN 0 901461 24 5 (Paperback)

Typeset by Impulse Phototypesetters Limited, St. Ives, Cornwall.
Printed by Headland Printers Limited, Bread Street, Penzance, Cornwall.

Contents

Publisher's Note

The Railway and Canal Historical Society wishes to express its gratitude to the South-East Wales Arts Association without whose financial support the publication of this book would not have been possible. The Society would also like to thank its friends in South Wales who have assisted in this publication.

Illustrations

Photographic sources: R—Gordon Rattenbury S—R. Derek Sach

Maps

The following standard symbols are used on all maps where possible:

■ Ironworks □ Colliery ▬▭▬▭▬ Incline(s)

△ Forge ● Limekiln(s) All scales are approximate.

I

Foreword

The effects of the period known as the Industrial Revolution were felt many years later in South Wales than in the Midlands and North of England. Until the mid-eighteenth century such industry as there was in the area consisted of widely scattered works founded where there was water-power available, with the additional requirement in the case of the small scale ironworks, where there was timber suitable for conversion to charcoal to use as fuel in the furnace.[1]

Although John Hanbury of Pontypool is reported to have used coal to re-heat wrought iron in preparation for rolling as early as 1704,[2] it was not until half a century later that the discovery, in 1709, by Abraham Darby, of Coalbrookdale, Salop, of a means of smelting iron using coal was introduced to the district. In 1757 John Maybery, who had a foundry at Powick, near Worcester, set up a small furnace at Hirwaun, near Aberdare, using anthracite coal as his fuel,[3] setting in train a series of events that resulted in South Wales being for a time the greatest iron producing district of the United Kingdom.

The northern rim of the Gwent and Glamorgan coalfield, from Abergavenny to Hirwaun, is a district abounding in good coking coal, iron ore, and limestone, all lying in close proximity. Lacking, however, was capital to exploit these resources; that had to wait until the English ironmasters had accumulated sufficient from the profits of their works to enable them to venture into this undeveloped field, with no real tradition of industrial working amongst the inhabitants. During the last quarter of the eighteenth century, a veritable rash of ironworks spread across the district.

The ironworks with which the Brecknock & Abergavenny Canal was concerned were all developed directly or indirectly with capital originating in England. Taking them in order from east to west, Blaenavon was founded by Thomas Hill, of Dennis, near Stafford, with two partners who came from North Worcestershire; Nantyglo, by Hill with some of the capital being supplied by the Bristol Quaker bankers, the Harfords; Beaufort and Clydach relied on capitalisation by the Kendall family from Shifnal, Shropshire, whilst the Ebbw Vale Works, although producing iron for many years in furnaces described by Lloyd[4] as primitive, had to wait until capital was injected by the Harfords before it could progress. The two remaining works with which the B&A was concerned, Sirhowy and Rhymney, were similarly financed, from London and Bristol respectively.

Abraham Darby came of a Dudley, Worcestershire, family. He left the district after serving an apprenticeship in Birmingham, and set up as a maker of malt mills in Bristol, a city strongly associated with the Society of Friends, where he met the Harford family, with whom he maintained contact after moving to Coalbrookdale, where his discovery was made.[5]

There were thus in the present county of Gwent a number of ironmasters emanating from districts where canals and tramroads were the accepted method of distribution of industrial products, or who had associations with those areas and who would know the benefits conferred by such transport. With capital provided, lack of suitable communications. was the great stumbling block besetting the early ironworks and preventing them from developing their full potential. It is not surprising that the ironmasters were behind the formation of both the Glamorganshire and Monmouthshire Canals, with their attendant tramroads.

The B&A, on the other hand, originated with the desire of the 'gentry' of Breconshire and north Monmouthsire to obtain the manufactured goods which were in the 1790s becoming available in areas with good communications, and to provide a means of exporting agricultural goods required to support the growing populations of the industrial districts. That it would provide a means of increasing the availability and decreasing the price of coal, particularly in Breconshire, was undoubtedly a major consideration.

Due to the mountainous nature of the country, the B&A is confined to the Usk valley, and for the greater part of its course is on a shelf cut into the hills to the south of the valley. For the first twenty-three miles from Pontymoil to Llangynidr, it maintains one level, and then rises forty-eight feet by means of five locks in half a mile. It resumes a level course for the next seven miles to Brynich, where it rises ten feet by one lock to reach the level of the terminus in Brecon, two miles further on.[6]

In order to reach the coal mining district, the B&A Act[7] authorised the construction of a " . . . Rail or Waggon Way or Stone Road from . . . the Canal at or near a place called Aberclydach . . . to Llangroiney Works . . . and from Aberclydach . . . to a certain place called Wain Dew . . . " the latter place being to the west of the present town of Brynmawr. Also authorised was a similar communication from a convenient place on the canal to the town of Abergavenny, which was never constructed. Possibly more important, it also authorised the construction of roads or railroads to any "work" within eight miles of the nearest point on the canal ("work" being defined as almost any form of manufacturing or extractive industry), should such be requested by the owner of the site concerned; should the canal company be unwilling to undertake the construction, the powers for the compulsory purchase of land granted to the canal could be made available to the owner of the "work".

It may be as well to recall how the "mileage" clauses came to be a regular feature of Acts of Parliament authorising canals in England as well as in Wales.

The Tyne and Tees valleys in north-east England were the districts in which the export trade in coal first developed, the mines, shallow in the main, being near the rivers, and convenient for loading the coal into ships. As the riparian mines became exhausted, or unworkable due to lack of suitable drainage, the collieries moved increasing distances from the rivers necessitating some form of transport to the shipping places. At first by pack animal or horse and cart and later by waggon-way,[8] this transport had frequently to cross land which was not in the same ownership as the collieries, and the landowner, naturally, would require recompense for the loss of use of his land. Often exorbitant sums were demanded for these "wayleaves", and by 1674, transport had become so great a part of the cost of coal that there was an attempt to get the government to lay down what should be considered a fair charge. Failing in this, the colliery owners presented a Bill in Parliament in 1796 with the same objective, but were again unsuccessful. Wayleaves remained a matter of great contention in the Tyne-Tees area (where no canals were ever constructed), until railways constructed under Parliamentary powers, with their rights of acquisition of land, superceded the privately owned waggonways.[9]

In order to avoid similar impositions on industry for communications with the canals which were being promoted in the eighteenth and nineteenth centuries, it was customary to include clauses in their Acts defining the distance within which it would be legitimate to construct roads, railroads or canals to connect with the authorised canal, without further recourse to Parliament. The distance varied in different Acts but was normally four, six or eight miles.

The engineer employed by the B&A was John Dadford, who was assisted in laying out the line of the canal by his brother, Thomas Dadford, Junior, and each produced his own ideas for submission to the shareholders.[10] John, however, was solely responsible for the line of the Clydach Railroad, from Glangrwyne to Wain Dew, and the branch to connect it with Edward Kendall's colliery at Gelli Felen. When John was appointed engineer to the Montgomeryshire Canal in 1794,[11] his brother was appointed to the B&A on a part-time basis to fit in with his work on the Monmouthshire Canal; in 1798, when his work there was finished he finally replaced John, who had gone to America in 1797.[12]

Thomas Dadford, Senior, had taught the brothers their trade, having wide experience of canal engineering in the Midlands before the family undertook work in South Wales. It would be surprising if, in the course of his employment on a part of the Trent & Mersey Canal in 1784[13] he had not heard, and possibly taken part in, discussions on the new railroad then being constructed between Froghall Basin and the limestone quarries at Caldon Low.[14] During his time with the Glamorganshire Canal he set out the route of the railroad to connect the Dowlais Ironworks with the canal at Merthyr Tydfil,[15] and it is probable that his advice on the type of track to be employed was followed. This closely resembles that used at Caldon Low, consisting of flat cast iron plates fixed to longitudinal timber baulks. In neither case is the precise gauge known, but it was probably in the order of three feet six inches or slightly narrower.[16]

For their lines in connection with the Monmouthshire and the B&A Canals, the Dadford brothers each adopted a three feet four inches gauge edge-railroad, using bar-rails of sufficient strength to enable the longitudinal wooden baulk to be dispensed with. Both experimented with cast iron tie-bars, or sleepers, but with no real success.

Writing in 1825, George Overton,[17] the engineer of the Penydarren and the Brinore Tramroads, was always careful to refer to lines laid with edge-rails as "railroads", and lines of plate-rails (laid with "L" section rails), as "tramroads"despite the fact that in his time the term "railway" had come into quite common use, and had been incorporated into the titles of several plateways in the district by their Acts of Parliament (e.g. Hay Railway and Llanvihangel Railway).

For the sake of clarity, the term "railway" has, as far as possible, been reserved for what is understood by the term in modern parlance. Occasionally, this has not been possible where quotations from documents or minute books have been made, as not all lawyers or company clerks observed the niceties as did Overton.

In the following pages, only tramroads and railroads constructed under the powers (including the "eight mile clause"), of the B&A Act of 1793 have been described, with the addition of the lines made by Bailey brothers of Nantyglo and Beaufort Ironworks, which were of necessity closely integrated with those of the B&A. All the lines leading to the canal, no matter by whom constructed, had under the Act, to be "publick", and open to all comers on payment of the appropriate toll or tonnage. In every case the freighter had to provide his own waggons and means of locomotion. In other words, their function was very similar to that of a turnpike road or canal, and apart from the fact that the traffic was carried on rails had very little in common with the modern railway, where track, vehicles and motive power are all the property of the owning company—a concept first exemplified on the Liverpool & Manchester Railway in 1830, but never applied to the horse-drawn lines in South Wales. In its early days the B&A owned both boats and waggons which it let out to freighters, but in later days it only owned sufficient vehicles for maintenance purposes.

In the course of research one factor, common to all the lines studied, has emerged—in every case the conception and building are fairly fully recorded, the day to day working very difficult or impossible to unearth, and the demise absolutely unrecorded. No attempt has been made, in most cases, to carry the story beyond 1865, when the B&A surrendered its independence to the Monmouthshire Canal.

It was found impossible to deal with the lines described in chronological order due to the number of co-incident events; the sections have, therefore, been arranged in geographical order, working from east to west.

Distances, weights and values have all been quoted in imperial units, as it was felt that contemporary units were better suited to an historical study.

In the lists of references the following abbreviations have been used:

D&C.	David & Charles Ltd., Dawlish.
GRO.	Gwent Record Office.
IRR.	*Industrial Railway Record,* the journal of the Industrial Railway Society.
NLOW.	National Library of Wales.
PRO.	Public Record Office, Kew.
RM.	*Railway Magazine.*

REFERENCES:

1. Rees, D. Morgan. *Mines, Mills & Furnaces.* H.M.S.O. 1969.
2. Rees, William. *Industry before the Industrial Revolution.* University of Wales Press, 1968. Vol. 1. p. 296.
3. Rees, D. Morgan. op. cit.
4. Lloyd, John. *Early History of the Old South Wales Ironworks, 1770–1840.* 1906.
5. Raistrick, Arthur. *Dynasty of Ironfounders.* Longman Green & Co. 1953.
6. Stevens, R. Alan. *Brecknock & Abergavenny and Monmouthshire Canals.* Towpath Guide No. 2. Goose & Sons, 1974.
7. 33 George III. xcvi. (1793).
8. Waggon-way was the name given to early railroads in north-east England.
9. Nef, J. U. *The Rise of the British Coal Industry.* George Routledge & Sons, 1932. Vol. 1. contains many references to wayleaves.
10. NLOW. Maybery 428.
11. Hadfield, Charles. *Canals of the West Midlands.* D&C. 1966. p. 189.
12. ibid. p. 190.
13. Hadfield, Charles. *Canals of the East Midlands.* D&C. 1966. p. 23.
14. Lewis, M. J. T. *Early Wooden Railways.* Routledge & Kegan Paul, 1970. p. 284.
15. Elsas, M. (Ed.). *Dowlais Iron Company's Letters, 1782–1860.* Glamorgan County Records Committee, 1960. p. 148.
16. ibid. pp. 148 and 171; RM. June 1937, Hollick, J. R. *The Caldon Low Tramways.* Both read in conjunction with IRR. No. 59. Lewis, M. J. T. *Steam on the Penydarren.*
17. Overton, George. *Description of the Faults and Dykes of the South Wales Mineral Basin.* 1825. Vol. 1. It appears that Vol. 2 was not written.

2

The Brecknock and Abergavenny Canal

Whilst in 1814, Walter Davies could write that the Brecknock and Abergavenny Canal was "... the only one as yet, in the district, which originated with the consumers; being undertaken with the laudable view of lowering the price of coal, lime etc. ... ",[1] it is obvious that there must have been more commercially minded people interested.

As originally suggested in 1792, the Abergavenny Canal, as it was then called, was to make a communication between the river Usk at its tidal limit, at Newbridge on Usk, to Gilwern, the nearest point to which coal from the Brynmawr area could easily be carried. It would thus have provided a competitive route to the sea with the Monmouthshire Canal Company's proposed lines to Blaenavon and Nantyglo from Newport.

In September, the scheme was widened with the suggestion that the canal should be continued to Brecon,[2] and it was probably the subscribers thus brought in that Davies had in mind; the subscribers resident in Monmouthshire would have been comparatively favourably placed in any case to obtain the commodities mentioned.

Scenting competition, and anxious to bring some of the possible trade to their own canal, the committee of the Monmouthshire Canal approached the subscribers with the suggestion that the line of the B&A should be amended to make a junction with the Pontnewyndd line of the MCC at Pontymoil instead of running into the Usk. By this means, the B&A would be saved some 220 feet of lockage, as the line of their canal at Llangynidr was on the level of the MCC at the proposed junction. As an inducement the B&A was offered a cash sum of £3,000, and the water to fill the long pound which would result, from Llangynidr down.[3] It is difficult to see how this was to be done, as with its heavy lockage the MCC was far more likely to draw water away from the junction.

On 16 October, the MCC committee ordered their engineer, Thomas Dadford, Junior, to make a survey of the proposed B&A and its railroad connections.[4] In this, Thomas appears to have collaborated with his brother John, and set to work immediately. At a meeting in Abergavenny on 8 November, attended by representatives of the Monmouthshire and Breconshire subscribers, it was unanimously resolved:

> "That the line from Brecon to Cwm Crawnon [near Llanddetty] being the same laid down as well in the Plan of Mr. Thomas as in the Plan of Mr. John Dadford be adopted. That the lower line being that of Mr. John Dadford from thence to within about two-hundred yards of the field between the Turn Pike Road and Mr. Waddington's House [near Llanover] be also adopted. That a Survey be taken from that Point by the easiest and most practical Line to join the Monmouthshire Canal near the Town of Pontypool."[5]

It was also decided to construct railways from Langroiney [Glangrwyne] Works to Gilwern and up both sides of the Clydach valley to Brynmawr and Beaufort Furnace.

The following Wednesday, the Breconshire subscribers met at Brecon, attended by four of those who had been present at Abergavenny, and decided that unless their stipulations (unfortunately unspecified) were met at the joint meeting to be held on the following day:

> "... the Breconshire subscribers divide from the Monmouthshire and take that part of the line within the County of Brecon to make separately and apply for an Act for that purpose. If that cannot be obtained the Brecon Line Subscribers at all events must insist on the Rail Way up the Clydach Valley or give an Opposition in Parliament."[6]

Here undoubtedly was the spirit of independence that led to the formation of separate committees, and the appointment of separate treasurers for the two counties, when the plans were eventually put into operation.

The finalised plans were submitted by Thomas Dadford, early in 1793 and Royal Assent given to the Act on 28 March.

The clauses in the Act with which a study of the tramroads and railroads constructed under its auspices is concerned are numbers 96 and 108. Section 96 authorised, in addition to the rail roads specifically mentioned in the general objectives of the company, the construction at the request of owners of "works" lying within eight miles of the canal or its ancillary lines, of "Stone Roads or Rail Ways," and extended the company's powers to acquire the necessary land. Should the company decline to make the line, after six months the applicant could acquire the right to purchase compulsorily the land required subject to negotiation, and make the line for himself. Should difficulties be placed in his way by landowners, the Act provided for a legal commission to be appointed to settle a fair price. Roads thus made were to be deemed "publick and open to all Persons", subject to the payment of tolls or tonnages at the rate applicable on the company's own lines. It will be gathered that these powers were exceedingly wide in their application.

Section 108 was concerned with interchange traffic between the B&A and the MCC, and eventually came into force in 1812, when the junction was made.

In essence, the MCC had been originated by the ironmasters and other industrialists to provide convenient, if not particularly cheap, conveyance to Newport. The B&A on the other hand, and in particular the Breconshire section, could be termed an agriculturally orientated concern, in spite of the fears of the MCC that coal from the Clydach valley might compete with that originating in the districts served by its canals and railroads.

The differing grounds of their foundation were reflected in the tonnage rates authorised in the respective Acts, viz:

	MCC per ton/ mile	B & A per ton[7] mile
Iron ore, lead ore, coal and coal products.	2½d.	2d.
Lime, limestone and quarry products, hay and straw.	1½d.	1d.
Iron and lead.	5d.	3d.
Timber and general merchandise.	5d.	4d.
Cattle, sheep and swine.	½d.	4d.
Tolls for passage of animals on the railroads:		
Horses, mules and asses.	1d. each.	1d. each.
Cows, horned and neat cattle.	½d. each.	2d. each.
Sheep, swine and calves.	Not quoted.	5d. per score.

Probably these rates show better than anything else the differences in the considerations of the two bodies of subscribers.

Whilst the MCC could see the danger of competition, the B&A could see that there was a possibility that, with the two canals making an end-on junction, tonnages on the MCC could force up the price of goods imported. Consequently, and possibly as part of the price for making the junction, section 108 was included in the B&A Act. Under this, "Coals, Goods, Merchandizes, or other Things" *originating or terminating** on the B&A were to be carried throughout at the current rate on the B&A, provided that this was not in excess of that being charged by the MCC.

In later years, this section was to cause the construction of several tramroads to the B&A, and agreements between the company and various ironmasters which were hotly contested by the MCC, who used as their justification the omission of the specific mention of iron from the clause. This resulted in the long run, in the two concerns engaging in bitter competition, instead of being complementary to each other, as had been thought would be the case at their inception.

The first assembly of the proprietors of the B&A was held on 16 May 1793, at which two committees were elected, one for each of the two counties concerned, and a treasurer for each was appointed; a somewhat clumsy arrangement, that could only delay decisions, and one which was abandoned in 1804, when a single treasurer and a committee of twelve was substituted.[8]

In addition to the canal, the Act authorised the construction of " . . . a Rail or Waggon Way or Stone Road . . . " from the canal at Aberclydach (Gilwern) past the Llanelly Furnace,[9] to Wain Dew, between Brynmawr and Beaufort; a similar way from Gilwern to Llangroiney (Glangrwyne) forge, and another line from a convenient place between Gilwern and Llanfoist to Abergavenny. The first two were built as the Clydach Rail Road, but the latter was never made, and Abergavenny had to await the coming of the Llanvihangel Railway in 1814 before having other than road communication with the B&A.

Bowing to the wishes of the Breconshire party, the assembly of 16 May 1793 ordered that the railroad from Glangrwyne to the collieries should be the first construction undertaken, and the decision was implemented by the committee on 14 June, when it was resolved " . . . that the part of the Rail Road from Langroiney to Cwm Gelli Velen Coal pits be immediately made."

Probably this line up the west side of the Clydach valley was opened in stages; it was in use by 15 October 1794, at least in part, as the assembly on the following day ordered that compensation of £5 5s. should be paid to William Powell in respect of his mare that had been killed "in consequence of a defect in the Rail Road"; and the same meeting ordered by-laws to be printed governing the use of the "Rail and Stone Roads belonging to the Company".[10] It is difficult, however, to assign a precise date to the opening, as once the utility of the railroad was realised, various extensions were authorised.

As yet, no attempt had been made to construct any part of their canal, and the committee of the MCC ordered their Chairman, Edward Kendall of Beaufort Ironworks, to attend the B&A meeting of 16 October to ascertain what steps had been taken to justify their acceptance of the £3,000 from that company.[11] Kendall was informed that " . . . the same will be done as soon as the Rail Roads are completed, and a proper engineer can be procured to conduct and superintend the same". Thomas Dadford, who attended the meeting in his capacity as a shareholder in the B&A, was ordered to mark out and stake the line of the canal. This order had to be ratified by the MCC committee on 20 October, as, officially, Dadford was still in their employ.

A year later, the MCC sent a deputation of three to attend the B&A assembly,[12] only to receive the answer " . . . that they may be assured it is the intention of the Proprietors to begin cutting the canal as soon as the necessary preparation for that purpose can be made."[13] and with that they had to be content. Eventually, the B&A employed Thomas Dadford to be their engineer "for the execution of the canal" as from 17 December 1795, on a part-time basis until his engagement with the MCC terminated at Christmas 1798.[14]

Various contracts for cutting the canal were let during the course of 1796, and many and varied were the persons to whom they were given. Local people suddenly became "canal cutters", even though their trades ranged from shop keepers to miners; amongst their number was a William MacDonald of Matlock, Derbyshire, but most were of local origin.[15] The contract for the embankment at Gilwern, leading to the Clydach aqueduct, went to Thomas Powell, Abergavenny, shop keeper, William Watkins of Llanvihangel Talyllyn, labourer, and William Parry of Tretower, gentleman; the construction of the actual aqueduct was entrusted to Benjamin James of Llanover, stone-mason.[16]

Cutting seems to have begun in early 1797,[17] and to have reached Llangynidr by December, when John Knowles, the company's Agent, was told to defer his progress report until the next meeting, when it was expected " . . . the works between Clydach and Llangunider Wharf will be finished."[18]

The contract for the tunnel at Ashford, near Talybont on Usk was let on 29 January 1798, and in July it was resolved to push on to Talybont as rapidly as possible. In October the assembly ordered that the canal should be finished to Brecon, and extended in the other direction to some point from which it would be convenient to make a road to Abergavenny. The section from Talybont to Brecon was opened on 24 December 1800.[19]

Meanwhile, on 7 March 1799, the Monmouthshire committee of the B&A had ordered the suspension of all contracts for canal construction in their Division, and recommended to their Breconshire counterparts that an engineer should be engaged to give his opinion on the advisability of constructing a railroad from Gilwern to Pontymoil in lieu of the canal; they suggested Benjamin Outram as a suitable person. Four days later the Breconshire Division committee resolved that the proposed survey should be left until after the assembly on 25 April. As it turned out, however, the decision to write to Outram requesting him to do the job was taken by the joint committees at a meeting held the day before the assembly.

Outram accepted the commission, and on 20 June the committee gave him a list of the heads under which they required his report. His opinion on the existing canal and railroads was asked; particularly he was to suggest what improvements he considered necessary, and to give his opinion on the manner in which Ashford tunnel was being made; to examine the proposed line of the canal towards Pontymoil; and to estimate the cost and comparable return of substituting a railroad for this section.

He worked exceedingly rapidly, and his report was ready by 1 July.[20] With regard to the canal, he considered that any trouble was entirely due to faulty masonry and puddling, and was easily rectified. On the subject of railroads, as was only to be expected, he recommended the adoption of " . . . Tram or Dram Roads as they are called in this Country [Wales] . . . ", and proposed the abandonment of the existing line on the west of the Clydach valley, and the substitution of one on the east, which would run by a gentle fall past Brynmawr as far as Daren Falen, where he proposed an incline to a point slightly to the south of the Clydach Ironworks, thence to be continued to terminate at the canal to the east of Clydach aqueduct, at Gilwern, and " . . . would there be about eight Yards above the Level of the Canal . . . " If the Clydach Ironworks traffic was to be carried on this line, instead of on the Llam-march Railroad (Section 8), it should be made double between the canal and Gelli Felen.

Amongst his suggestions was that a tramroad should be constructed from the end of the former at Gilwern, to Abergavenny using the line later taken by the Crickhowell-Abergavenny Turnpike (now B4246) and suggested that, for local deliveries in Abergavenny, the waggons ". . . should not be unloaded at the Wharf, but each Waggon, with the help of a crane, should be set upon a pair of wheels, on which it might be drawn to any part of the Town".

He approved of the proposed canal line to Pontymoil, but considered that there might be trouble from "open strata". As to continuing from Gilwern by rail, he again suggested a tramroad, which he thought could be made for £19,000. It should be carried above the line of the canal to obtain the advantage of a gentle fall the whole way to Pontymoil. Under this heading, he thought that such a tramroad would facilitate the construction of the canal at a later date.

Ashford tunnel he considered should be made by "cut and cover", with the earth over the arch well pounded to render it impervious to surface drainage. It is doubtful if this recommendation was followed, as on 24 January 1803, Jonathan Dixon of Wenallt Farm, which spanned the canal at this point, was granted an annual payment of £5 5s. to refrain from watering his lands near the tunnel, a payment which was continued to his widow.[21]

It is unfortunate that the committee minute book is not available covering the period when Outram's report must have been under discussion. Discussion there undoubtedly was, but in the end no alteration was made to the railroads, and the proposed tramroad to Pontymoil did not materialise.

With the canal completed from Gilwern to Brecon, the Breconshire proprietors had attained their objective, but the Monmouthshire committee could see that it was essential to complete the canal to the MCC if it was to pay its way.

Under its Act, the B&A was authorised to raise a capital of £100,000 by subscription, and a further £50,000, if needed, by subscription or borrowing. The assembly on 30 April 1801 learned that expenditure to date had been £120,000, and in order to complete the canal a further £30,000 would be required, which it was proposed to obtain by increasing the value of the £100 shares to £150, thus obliging all fully paid-up shareholders to contribute a further £30 per share.

A special assembly was held on 12 March 1802 to discuss the expediency of continuing the canal to meet the MCC, and to consider going to Parliament for an increase in authorised capital. Included in the agenda was a suggestion that powers should be sought to make a new ". . . Dram Road up the South side of the Clydach valley . . . and for power to discontinue the present Road . . . " as proposed by Outram in 1799.[22]

It appears that there was considerable opposition to the latter proposal, and even Edward Frere, who greatly favoured the continuation of the canal, was opposed to the alteration of the railroad, in spite of the apparent advantages to the Clydach Ironworks. As he wrote to John Powell, the Brecon solicitor, the meeting had:

> ". . . confirm'd me in the opinion I had formed that the very resolution to proceed would materially benefit our concern . . . "[23]

The more progressive members had their way, and Thomas Cartwright, of the Worcester & Birmingham Canal, was engaged to re-survey the line to Pontymoil, and to report on the parts already constructed.[24] On 21 October the committe was authorised to invite tenders for the construction of two miles of canal from Gilwern to Govilon.

During the first half of 1802, the B&A found itself closely involved in the Bill that the MCC was endeavouring to get through Parliament. This, *inter alia* provided for the establishment of the Sirhowy Tramroad Co., and contained three clauses which could have affected the B&A had they become law. These would have removed the MCC's freedom to reduce its charges, would have enabled it to remove any of its lines of communication without Parliamentary authority, and (most disliked of all), it would have been given an absolute monopoly of all transport in the area it served by compelling freighters to make compensation to the MCC for all goods carried by any other route.[25] The last mentioned would have nullified the eight mile clause in the B&A Act. Strong opposition was registered by the Breconshire committee, who urged their Monmouthshire colleagues to join them in petitioning Parliament to have the clauses expunged,[26] and Richard Blakemore, for all his shareholding in the MCC, actively lobbied Members of both Houses of Parliament to oppose them with some fair measure of success.[27] The first two clauses were withdrawn at a meeting of the MCC proprietors in early April[28] but the last was retained until after a meeting attended by representatives of the B&A in London on 26 April,[29] whilst the Bill was in committee in the House.

It was, however, money and not Parliamentary Bills that was delaying the B&A's advance to Pontymoil, and on 28 April 1803, it was resolved to apply to Parliament for authority to raise a further £80,000 and the necessary Act was obtained on 3 May 1804 (44 Geo. III xxix).

Meanwhile, Cartwright had been given the contract to complete the two miles of canal towards Govilon.[30] Presumably, he must have sub-contracted the work, as he was employed by the W&B until 1809.[31] There had also been a change in the executive, William and John Powell, the Brecon solicitors resigning their position as joint-Clerks to the B&A on 26 April 1804. In their stead, B. A. Griffiths, formerly the Agent was appointed Clerk, a position he held until his death in 1827.[32]

By July 1807, the MCC was again becoming anxious about the fate of the £3,000 that had been paid over in 1794,[33] and forwarded a copy of their assembly minutes on the point, which was considered by the B&A committee on 20 July. In reply it was stated that the assembly had already authorised the construction of the "missing-link", and that as soon as money was available it would be made. This did not satisfy the MCC, and three of their committee attended the B&A assembly in October, presenting a letter to the B&A proprietors demanding to know what compensation was likely to be paid to the MCC for losses they may have sustained due to the non-completion of the B&A.[34] A suitable reply was drafted at the meeting, pointing out that the canal would be completed as soon as possible; that there had been no avoidable delay, and "... that the last Act obtained by this Company will best explain the situation of this concern in that respect". Again, the MCC was dissatisfied, and at their general meeting on 4 November, it was decided to seek counsel's opinion as to their case for compensation.[35]

It would seem that the B&A had difficulty in raising money under its Act of 1804, and was still short of the capital to continue digging. Better prospects were around the corner, however, and the assembly of 18 May 1809, heard that Richard Crawshay, of Cyfarthfa Ironworks, Merthyr Tydfil, was prepared to lend the company £30,000 for twenty years at five per cent interest, secured by a mortgage of the canal tolls. In August it was agreed that Crawshay should be given six promissory notes of £5,000 each, payable in 1829, rendering his security negotiable. Another large loan was received from Lady Fenn, who was introduced to the company by George Frere, brother of Edward Frere of Clydach Ironworks. This was for £8,100 secured by twenty-seven notes, each of £300.[36]

Once it had been ascertained that Crawshay really intended to make the loan, things must have looked a lot brighter, and at the beginning of August 1809, William Crosley, hitherto engineer to the Rochdale Canal was engaged at a salary of £500 a year to complete the canal within three years.[37] He started work immediately, and in conjunction with John Hodgkinson re-surveyed the proposed line in time to report to the committee on 19 August. For his share of this work, Hodgkinson was voted a gratuity of £21. On 18 October the contract to cut 4,560 yards from Mamhilad to the junction with the MCC was let to William Seed, who undertook to complete the work by the beginning of May 1812. He was better than his word—the canal was officially opened to Pontymoil on 7 February 1812.[38]

As might be expected, the decision to complete the canal brought requests for the construction of tramroads to it; some were rejected, but the applicants were given permission to build them for themselves utilising the B&A powers; a number were rejected out of hand including one from Hugh Powell of Llanvihangel Court, for a line to Llanvihangel Crucorney—a line for which an Act was obtained later. Another tramroad constructed under its own Act, the Hay Railway, from Brecon to Eardisley, was conceived during this period. Under both Acts the B&A accepted restrictions on the powers of the eight mile clause, removing their right to make tramroads which might have competed.

A line which was sanctioned by the B&A was that requested by Bailey & Wayne of Nantyglo Ironworks, from their works to the Clydach Rail Road, which the B&A undertook to make for them, charging a nominal rent of £1 1s. annually for its use. Having got so far, on 14 November 1811, Bailey & Wayne proposed that their products should be conveyed on the Clydach Rail Road to the canal free, on condition that all their output was sent down the B&A, a proposition that appealed to the canal company. Further agreements were made in March 1817 with Kendall & Bevan of Beaufort Ironworks, and Harford Crocker & Co. of Ebbw Vale, under which the B&A would connect the Clydach Rail Road to their respective works, and in return for all their produce being carried on the B&A, the canal company would reduce its tonnage on iron from 3d. to 2d. per ton per mile.[40] This, of course, meant that the MCC would be obliged to reduce their tonnages to the same figure.

The MCC struck back on 6 May 1817, ordering that Beaufort and Ebbw Vale Ironworks should be charged 5d. per ton per mile, the maximum authorised under its Act, but that Nantyglo and Clydach Ironworks should continue to be charged 3d. per ton per mile, as those works had to use the B&A in any case.[41] Harford Crocker & Co. informed the B&A of the state of things, and the latter decided at their June meeting that although not justified in attempting to adjudicate between the MCC and a freighter, they considered that they should fight any attempt to interfere in their internal management. On 16 January 1818, the MCC decided to test the matter in the courts. The freighters continued to send their iron down the canals, but refused to pay any tonnages to the MCC, whose income suffered a noticeable drop, and by May 1818 their Clerk was instructed to ask the Ebbw Vale Co. to pay the undisputed part of their tonnages.

The ironmasters played a delaying game, and in July 1818, Kendall & Bevan managed to have the case adjourned to the following Spring Assizes. In their search for evidence the MCC instructed their solicitor to examine the books of the B&A to ascertain that the meeting at which the reduction in tonnage had been passed was legally constituted to come to a decision on such matters.[42]

Eventually, on 3 November 1819, the MCC issued an appeal to all the ironmasters that they should pay tonnages then overdue at 2d. per ton per mile "until the pending question is determined", and in December 1820 they again appealed for payment of 2d. until the matter should be settled. In vain, the MCC reduced their tonnage to 3d. per ton,[43] but the reduction was insufficient to tempt the ironmasters, as the B&A route was still cheaper by about 3d. per ton on the whole distance.

It is possible that the MCC offered some other form of inducement to Harford, Crocker & Co., who advised the B&A on 19 October 1819 that they intended to withdraw from the agreement at the end of a period of six months.[44] It might have been to their advantage had they held on a little longer, as in June 1821 judgement was given against the MCC, who had to refund tonnages previously overcharged.

The B&A gave a wholehearted welcome to the prospect of the Hay and Llanvihangel Railways bringing traffic to their undertaking, but had qualms when in 1812, the Hay proposed to amend their Act of 1811 to permit the tramroad to terminate at Brynich, two miles short of Brecon Wharf, their originally intended terminus.[45] The B&A despatched their solicitors to London to ensure that this amendment did not go through, in which they were successful. Had the Hay proprietors realised it, they could have had termini at both places by proceeding under the eight mile clause. The canal committee foresaw this, and alarmed at the possibility that another time they might not be let off the hook so easily, on 26 June 1812 decided to recommend that application be made to Parliament for the abolition of that part of the eight mile clause permitting anyone other than the canal company to construct tramroads.

In October, Benjamin Hall of Rhymney Ironworks, applied for a tramroad from his quarries at Trefil to the canal at Talybont on Usk[46] a line which later became the Brinore Tramroad. Investigation having shown that the application was covered by the eight mile clause, the committee confirmed their recommendation of 26 June, and a special assembly on 11 December put the wheels in motion. A copy of the minutes of the special meeting was sent to Hall on 21 December, by William Powell, the B&A solicitor, together with a letter informing him that if the Bill was passed in the form envisaged, not only would the powers of the owners of works be cut, but that the B&A, itself, would be powerless to build the tramroad of his application—the B&A's Parliamentary agents insisting that the one-sided arrangement proposed by the company stood no chance of success, and advising that the clause should be expunged in its entirety.

Naturally, the people of Brecon saw in the B&A application to Parliament an attempt to prevent coal being shipped at Talybont, leaving them at the mercy of the Brecknock Boat Co. who supplied coal from the Clydach valley, shipped at Gilwern. This meant that the BBCo. had to include in its prices at Brecon, twelve miles more canal tonnages than would have been the case had Hall's tramroad been made.

At a meeting held on 12 January 1813, in Brecon a motion was passed that "... the County of Brecon may take such measures as to them seem proper for the purpose of disproving the proceedings of the Canal Proprietors",[47] and in answer to the B&A's contention that they would lose considerably in tonnages should further tramroads be constructed, Samuel Homfray of Tredegar Ironworks, made the following offer:

> "The Canal Company complain that the tonnages will be lowered . . . if the Eight Mile Clause is continued and new Roads are made under it. In order to remove this I am ready to take a lease on the Canal for twenty-one years at the present income with an additional Hundred Pounds per annum to it."[48]

Another special assembly was called for 1 February and with the notice convening it a circular letter signed by Griffiths was sent, playing largely on the cupidity of the proprietors, which had the desired effect, and a motion to rescind the previous minute was lost by 612 votes to 170,[49] and there was nothing to stop the Parliamentary application proceeding. The Bill came up for its second reading in the Commons on 5 March, and on the opposition of Sir Charles Morgan, who had large estates in the Brecon area, was defeated by 162 votes to 93.[50]

There was criticism when the cost of the proceedings, £647 17s., became known, and Powell took up the question with the Parliamentary counsel, who replied that had the Bill been presented in its original form, it would have stood no chance of success, and the alterations at the last moment had increased the costs. With that the proprietors had to be content.[51]

The way was now open for the construction of further tramroads under the eight mile clause, and in the following ten years the Brinore (1815), Llangattock (1815), Watton (1816), Bailey's (1821), and Hill's (1823-4) were all made; the Watton and the Llangattock under the aegis of the B&A. One which would have benefitted the people on the line of the Hay Railway was not implemented, however, when in 1821, George Overton the engineer of the Brinore, drew plans for a tramroad to connect that line to the Hay at Talyllyn; a scheme which would have obviated seven and a quarter miles of canal and three and three-quarter miles on the Hay at the expense of four miles of new tramroad, with a consequent reduction in transport costs.

It was only cheaper carriage that ensured the use of the B&A by the ironmasters; in point of distance from the works to Newport, and consequently in journey time, everything was in favour of the MCC. Moreover, as far as Beaufort, Ebbw Vale and Nantyglo works were concerned, after 1828, the MCC offered the advantage of through tramroad communication with Newport, avoiding the expense of transhipment into canal boats. Clydach was the only works irrevocably tied to the B&A.

In May 1828, the MCC reduced its tonnage on iron to 3d. per ton per mile and with the completion of its tramroad to Nantyglo Gate (and the extension to their works by the Baileys) in the July, the outlook for the B&A was somewhat gloomy. In a pessimistic mood the B&A called a joint meeting with the ironmasters for 27 August, which resulted in a circular letter to the proprietors advising them that unless tonnages were further reduced, the iron freighters would be unable to continue to use the canal. The assembly on 16 October ordered a reduction from 2d. to 1½d. per ton per mile on iron, and in August 1830 the tonnage was again reduced, this time to 1d. per ton.

The MCC made a two-fold response. In July 1829, they authorised the loading of waggons to be increased to three and a half tons each, and in March 1830 it was decided that after profits sufficient to pay a dividend of £6 per share had been appropriated, any surplus should be divided amongst the freighters proportionally to their expenditure on tonnages. In May 1830, the freighters received a rebate of thirteen per cent of what they had paid.

With the two canals forming a continuous route, it was perhaps inevitable that mistakes made by one should affect the other. Such a case occurred on 3 January 1829, when the B&A drew the water out of their canal at Pontymoil without putting down stop-planks at the junction which "caused the trade of the Monmouthshire to be interrupted for two days".[54] No instructions were given, however, to ensure that the B&A was aware that the MCC intended to let the water out of their canal from 27 June to 8 July 1830 during which time all through traffic from the B&A would, of course, be stopped.[55]

As early as 1818, Harford & Co. had acquired a controlling interest in the Sirhowy Ironworks and made a tramroad to it from their Ebbw Vale Works, running parallel to the MCC's Rassa Railroad. The MCC regarded this as an attempt to breach what they considered was their rightful monopoly of transport in the district. Through the years numerous complaints were made to the Harfords that they were blocking the traffic on the Rassa, either by depositing rubbish on it, or by removing rails to enable them to make branches across it from their private tramroad to collieries they worked at Hirgan and Penmarc, close to the Sirhowy Works.

By 1828 it is probable that the B&A's Clydach Rail Road, which had been used occasionally by Harfords to reach the B&A at Gilwern, was in need of repair, and Harfords were seeking a fresh outlet, to avoid using the MCC's Beaufort Tramroad, as they were already in dispute with the MCC owing to something more than a suspicion that they were sending iron from Ebbw Vale to Sirhowy, and then down the Sirhowy Tramroad at a cheaper rate than was available from the MCC, contrary to the Act of 1802.

On 27 October, Harfords applied to the Brinore Tramroad Co. for permission to construct a tramroad to join that line. The Brinore were not anxious for the honour, and appointed one of their committee to ascertain if there would be sufficient profit in it to justify their taking on the extra traffic.[56]

The Brinore's decision was evidently unfavourable, as on 20 October 1831, Harfords notified the B&A that it was their intention to construct a railroad from the fifth lock on the canal at Llangynidr to their colliery and iron mines at Penmarc. The minute book does not describe the proposed route, but the *Hereford Journal* of 7 November reported that it was to lead past the Trefil Quarries, and down Dyffryn Crawnon to the canal within half a mile of Llangynidr.

The B&A took its time to consider the scheme, and it was not until 26 January 1832 that they asked Harfords for the plans, and in April it was decided to recommend that it should be constructed, should the proposition come within the terms of the eight mile clause.[57] However, at the assembly on 26 April the decision was put off until the end of the following July, when the committee should call a special assembly. Why the iron company was prepared to accept the delay is revealed by the Brinore minutes for 5 May, when Summers Harford attended their meeting having since his previous application acquired one share in that company, doubtless intending that it should be a lever to get them to bow to his wishes with regard to using the Brinore. He was informed that it would be necessary for him to submit a written application, which he refused to do, saying that he had already written to their solicitor, who denied all knowledge. Nothing more was heard of the application.

By June the scene of action was transferred to the MCC, whose committee was informed on the 18th that Harfords had again taken up rails on the Rassa line, intending to extend their private line across it. Nine days later it was minuted that notices had appeared in the *Cambrian* that it was intended to construct a tramroad from Sirhowy to the B&A, and it was decided that any such move should be opposed by all possible means, and that both Harfords and the B&A should receive a copy of the minute, and Harfords were further warned that if any rails were laid on MCC property, they would be proceeded against for wilful trespass.

Doubts were expressed at the B&A special assembly on 27 July 1832 as to their ability under the Act to comply with Harford's wishes, and Summers and Charles Harford, who attended, concurred with the suggestion that counsel's opinion should be taken. Undeterred, Harfords applied for yet more lines in January 1833, all to branch from that originally requested. They were told that the B&A intended to retain their Parliamentary powers in respect of the original line, and rejected the branches subject to the confirmation of the next assembly.

At the assembly the full extent of Harford's requirements were revealed to consist of the originally proposed tramroad from Penmarc to Llangynidr, with branches to Trevil Machine—on the Trevil Rail Road; to a quarry on Llangynidr Mountain; and to the source of the Crawnon Brook, where their successors, Richard Thomas & Co. Ltd., opened the Chwar yr Hendre quarries in 1938. There was a further request for a line from the Hendre, near Shop Row, Beaufort to the canal at Gilwern. This latter being good evidence that the original Clydach Rail Road was, at least in part, impassable. The assembly concluded that they should exercise their powers in respect of the original application, and ordered the Clerk to make the necessary applications to the landowners, but should refuse to make the other lines.

This assembly must have had an unusually busy time, as it also considered and rejected an application from George Brewer of Coalbrook Vale Ironworks for a tramroad from near Nantyglo to Llangattock. Another matter under discussion was the possibility of co-operation instead of competition with the MCC.

James Peirce, who was now Clerk to the B&A, was instructed to write to the MCC enquiring if they had any objection to the new tramroad crossing their Rassa line, but the MCC was implacably against the idea. Possibly in order to test the strength of the reaction, on 25 July the B&A committee ordered Peirce to commence building the tramroad starting at the Rassa near Penmarc. The timing of this instruction seems to be remarkably inept, as the same meeting appointed a deputation to meet the MCC on the suggested co-operation!

The deputations met on 6 September 1833 and on the advice of solicitors decided that amalgamation of the two companies was impractical due to the wording of both Acts, and a further meeting on 6 November concluded that further discussion was unlikely to produce a result and the talks should be abandoned. The probable reason for the lack of progress had been revealed at the MCC committee on 8 October, when there were complaints of the amount of traffic the B&A was taking from them, estimated at 39,000 tons annually of which " . . . this Company might fairly and did undoubtedly when their works were formed expect the transport . . . " and which was reckoned as depriving the shareholders of £2 annually of their dividend. From the viewpoint of the MCC this would hardly have formed a suitable background for amicable discussions.

Peirce had not started to make the tramroad for the Harfords, and the B&A assembly of 24 April 1834 received a letter from Harford's solicitors in Bristol, Isaac Cooke & Sons, demanding that the various tramroads should be made forthwith. It was decided that the letter should "be received", but no action was ordered. Action was forced upon them, however, when the B&A's solicitor called a special committee meeting for 9 July to consider Harfords application to the Courts for a *writ of mandamus* to force them to make the tramroads mentioned in Cooke & Sons' letter.

It was resolved to defend the action, and a prerequisite was the provision of evidence as to why the tramroads had not been made. The MCC minute books describe events, with no necessity for comment.

On 4 November the following letter was received by the MCC committee:

Trevil Machine
October 25th 1834

Thomas Cooke Esq.

Sir,

I beg leave to inform you that Mr. Edward Davies of Llangattock [a surveyor] and Mr. Pierce of Llanelli came up yesterday to the Trevil Machine with a wagon-load of planks and barrows a half dozen or more men to set off the new Tram Road from Sirhowy to Langunnider and they have begun to form the Road just below the second cross but they have not come to our boundary. Sir, William Jones begs you would you be pleased to send him a line by the return of post to let him know how you would wish him to act and he will wait at the Greyhound in Tredegar tomorrow night for your answer.

Your humble Servant, Thomas Philips.

Their reaction was swift and to the point, and in a letter to Peirce it was pointed out that it was apparently intended to cross the Rassa at a place where Harfords had already removed the rails, and an action against them was pending, and concluded with a stern warning that should Peirce or any workman in the B&A's employ trespass on MCC property, he and they would undoubtedly by prosecuted.[59]

With such evidence of their inability to make the tramroad, it was small wonder that on 18 May 1835 the B&A solicitor was able to report that the action had been dismissed with costs against the plaintiffs. He also brought a considerable amount of correspondence he had had with Gabb & Secretan, Harfords' local solicitors in Abergavenny, demanding an explicit answer as to the intentions of the company in respect of the tramroads. On 9 June, a special assembly confirmed the decision not to let Harfords make the lines for themselves.

The tramroad to Llangynidr is heard of no more, but in April 1839 Harfords requested the B&A to put the railroad from the Rassa Rail Road to the canal into full working order. The B&A made no move, and on 3 July, Isaac Cooke & Sons wrote on their behalf requesting the canal company "... to relay the Roads sufficient for the purpose of the trade thereon ..." The letter was considered by the committee on 25 July, who instructed their solicitors to reply:

> "If Messrs. Harford & Co. will undertake for the space of seven years to carry a remunerative trade along the Rail Road they seek to be re-laid the Company will take their application into consideration. The Company consider otherwise the Road to be utterly useless."

Probably the B&A had heard whispers of the state of the finances of Harfords, who were bankrupt in early 1843. The B&A does not appear to have lost any money, but the MCC, despite having seized many waggons and a locomotive, had to accept compensation of 14s. in the pound in respect of overdue tonnages.[60]

Following the success of the Liverpool & Manchester Railway in 1830, few parts of the British Isles were spared the spate of schemes, mostly abortive, that flooded the country. The Usk valley was the objective of several proposed routes.

By December 1836, three such schemes had been brought to the notice of the B&A committee; one, from the east side of the Usk at Newport to Pontypool and Abergavenny did not even deposit plans. Plans were deposited, however, for the Union Railway (from Gloucester to Fishguard), which would have crossed the canal several times, and the Brecon & Merthyr Tydfil,* which would have paralleled it for several miles, and both were opposed by the B&A.[61] Nothing more was heard of railways in the district until the mania year of 1845.

On 30 January 1845, the committee decided to oppose the MCC's intention to apply to Parliament for a railway from Newport to Pontypool, but the MCC obtained their Act. The B&A committee decided in September after being approached by the Newport, Abergavenny & Hereford and the Welsh Midland Railways, both of which proposed to use the line of the canal as part of their schemes, that if the concern could be disposed of to a railway company, it would save what was left of its capital.[62] It was finally decided to recommend to the assembly that the offer of the WMR of £182,500, sufficient to pay off the accumulated debt and give £146 for each £150 canal share should be accepted.[63] The assembly ratified the decision, and the WMR paid ten per cent of the purchase money as a deposit, out of which the B&A repaid a debenture of £6,000.[64] Unfortunately for the B&A, the WMR did not obtain its Act, and the assembly of 30 April 1846 was faced with a demand for the repayment of the deposit. In this awkward situation, the B&A could only temporise, and called a special assembly for 30 June, which decided to inform the WMR that they did not propose to repay until such time as the WMR had finally abandoned its proposals. It was not until 15 October that it was decided to pay up, which they were enabled to do in January 1847, when Joseph Bailey lent them £6,000.[65] It may be that Bailey had twinges of conscience, being chairman of the WMR.[66]

Two years elapsed before the next attempt to improve their fortunes, when the May 1849 assembly voted £50 to have a survey made of the possibility of adapting the canal line to a

* Not to be confused with the later Brecon & Merthyr Tydfil Junction Railway of the 1860s, commonly known as the B. & M.

railway, " . . . pointing out the places where it would be necessary to deviate from the present curves . . . "[67] The October assembly passed a vote of thanks to Crawshay Bailey for his trouble in obtaining the survey, and appointed a sub-committee to inspect the plans. Nothing more was heard of this project, however.

After a further two years, a deputation from the town authorities of Brecon attended the assembly of 24 April 1851, and asked if the B&A would be willing to sell the canal between Brecon and Abergavenny for conversion to a railway. It was decided to call a special assembly to discuss the matter, and a price of £105,750 was suggested. The special meeting was held on 10 June, and agreed to sell subject to the Breconshire Railway, as the promoters called their scheme, obtaining its Act. On 29 January 1852, as no effort had been made in that direction, the promoters were advised that the deal was considered to be null and void.

Under the MCC Act of 1845 for a railway from Newport to Pontypool, their canal from Pontymoil to Pontnewynydd was to be abandoned. Due to the financial stringency following the collapse of the "railway mania", a further Act was necessary in 1848, extending the time for completion of the railway. This forbade the closure of the canal in Pontypool until the whole of the railway from Newport should be completed; authorised an increase in their share capital, and changed the name of the company to the Monmouthshire Railway & Canal Co.[68] With money still scarce, little progress had been made on the N&P by 1851, and in July it was decided to go to Parliament for a further extension of time, and *inter alia,* to obtain powers to close the whole Pontypool line of canal for a period of six weeks to facilitate construction work on the railway.[69] Naturally, such a closure would mean that all traffic from the B&A would be prevented from reaching Newport, as indeed, would that from the Blaenavon Works, whether it was sent via the B&A or down the Blaenavon Rail Road to Pontnewynydd. The point was raised in the B&A committee on 19 November 1851, and the solicitor was instructed to demand of the MRCC an explanation of the clauses authorising them to " . . . stop, divert and alter their or any other canal . . . "

At the MRCC assembly on 31 March 1852, a Mr. Williams, who described himself as a "director" of the B&A, stated that he did not think that company would insist on the MRCC keeping open the canal above Pontymoil, but that any closure lower down for six or even four weeks, would be considered objectionable; he would, nonetheless, endeavour to obtain modifications in the B&A's petition against the Bill.[70] Joseph Bailey reported to the B&A committee on 3 April that at a meeting with the MRCC, that he had attended in his capacity as a trader on both canals, the MRCC had undertaken to convey all cargoes from the lowest point to which navigation was possible to the docks without any extra cost to the freighters.[71]

On 16 April, a meeting took place between delegations from the two companies, at which it was agreed that, subject to the B&A withdrawing its petition against the Bill, the MRCC would make good any deficiency in the B&A's income in respect of that part of their trade which was also carried on the MRCC's canal whilst the water was let out of the Pontypool branch, and on the representations of the Nantyglo and Clydach iron companies, the MRCC undertook to convey all cargoes over the break, " . . . keeping all Boat loads separate if requested . . . ", at no extra expense.[72] Similar concessions were made to the traders on the Crumlin branch and the coal traders at a meeting on 29 April.

Royal Assent was given to the MRCC Act on 30 June 1852, and in response to their enquiry, they were informed by the B&A that the latter had no knowledge of any scheme to make a railway to Brecon that would render their canal obsolete, and the MRCC hope that they would not need to re-open the Pontypool canal was dashed.[73]

Early in 1853, the London & North Western Railway backed a scheme in the name of the Newport, Abergavenny & Hereford Railway for a line from Abergavenny to Brecon, and approaches were made by the B&A to the LNW in the hope of selling the canal to them, but on learning that it was not intended to follow the canal line, it was decided to petition against the proposal[74] which was, however, withdrawn due to Parliamentary opposition to the spread of LNW influence over the NAH.[75]

A year later, the B&A itself authorised a survey for a railway between the two towns, and a subscription was ordered to be opened.[76] On 19 October 1854, the committee appointed to look into the project reported that as the MRCC was rapidly approaching the north Monmouthshire ironworks with its conversion of its tramroads to railways, they considered it advisable that any proposed railway should be made for the whole distance from Brecon to Pontypool. A motion that this should be done was defeated on a vote.

Another bid was made for the B&A to the assembly of 5 September 1856, when McCormick Brassey & Co., the railway contractors, offered £122,000 for the canal line from Brecon to Abergavenny, payment to be made in shares and debentures in the new concern. The offer was accepted, despite an amendment that only cash offers should be entertained. Once again, no action followed.

The assembly of 21 October 1858 was attended by two representatives of Messrs. Tootal, Johnson & Simpson (probably acting on behalf of the NAH), who offered a guaranteed rent charge of £4 p.a. per share and an undertaking to purchase the shares for £80 each within the next four years, and to take over the B&A debt of £36,450, in return for a lease of the canal from Gilwern to Brecon, leaving the remaining portion from Gilwern to Pontymoil in the company's hands. On 29 October a draft provisional agreement was signed by both parties.

This new company, the Breconshire Railway & Canal Co.,* also made an agreement with the embryo Brecon & Merthyr Tydfil Junction Railway, whose Bill had been published on 9 November 1858, that the B&M should have access to Brecon over their rails from Talybont.[77] The failure of the BR&CC to push forward its Bill was announced to the B&A assembly on 28 April 1859. This left the B&M, who had deleted the Talybont to Brecon section from its Bill, with the necessity of obtaining their own access, which was eventually done with the acquiescence of the Hereford, Hay & Brecon Railway, who had purchased the Hay Railway.[78]

No further attempts to sell the canal were made, and the company seemed to resign itself to obtaining what concessions it could from those railways which did materialise. Meanwhile, the iron trade was drawn steadily away from them by the progress of the MRCC railways to the works, and as had been the case in the early days of the canal, they were probably entirely reliant on what trade could be provided in coal and lime by the Brecknock Boat Co.

In 1860, the B&A obtained an undertaking from the B&M solicitor that they would provide a siding at Talybont to serve the canal, and after lodging a petition against the Hay Railway (Tramroad) Bill, under which the HH&B was to acquire the Hay, the B&A was granted a reprieve in that the HH&B had to undertake to keep open the tramroad between the canal at Brecon and a point three furlongs short of Hay Bridge until such time as their railway was completed.[79] In 1861, permission was given to the B&M to construct a bridge over the canal at Talybont, and the canal company undertook to let out the water from the canal for seventy-two hours to facilitate the operation.

* Promoted jointly by the NAH and some members of the B&A committee.

In 1861, with the withdrawal of virtually the whole of the iron carrying trade by works which could use the MRCC's new railways, it came as a bitter blow that the Clydach Iron Co., who were entirely reliant on the B&A for their carriage, were unable to pay their tonnages, amounting to £1,650. They offered to pay this at the rate of £100 per month, but were unable to manage this comparatively small amount, and judgement was obtained by the B&A for the full sum in July. The works was sold in 1863, and the B&A paid in full by the trustees in bankruptcy, but not before it had declared its inability to pay any dividend in 1862 due to the deficiency.

The advent of the proposed Vale of Crickhowell Railway, which obtained its Act in 1862 for a line from Crickhowell to Abergavenny and a further Act, in the following year, to continue to Talybont, forced the B&A to the conclusion that it had very little future unless it could shelter under the aegis of the MRCC. The promised siding from the B&M at Talybont had not materialised, nor had the company been able to pursuade the joint HH&B and B&M to connect the canal to the railway at Brecon.

The Clydach Rail Road between Brynmawr and Beaufort and Ebbw Vale had been virtually unused since 1839, and the site had been the scene of activity by squatters who had erected houses or made gardens, and in 1863-4 parts were officially let to the people who had seized occupation.

At the end of 1864, an agreement was reached with the MRCC under which that company should give £25 per share on the 990 remaining shares in the B&A, and make themselves responsible for the debenture debt of £36,450. No expenses for the transfer were to devolve onto the B&A, and the MRCC undertook that no part of the B&A canal or its attendant tramroads should be closed without Parliamentary authority.[80] The assembly accepted the offer, and the MRCC obtained an authorising Act (28-9 Vict. cclxxxi), which came into effect in September 1865.

The final dissolution of the company did not take place until June 1867, and in the interim, the railroad to the west of Clydach Bridge, Brynmawr, was sold to J. & C. Bailey for £300.[81]

There is, perhaps, a degree of irony in the fact that, whilst no railway was built in the Usk valley between Abergavenny and Brecon, the B&A is still largely navigable, but the MRCC's canal has to a large extent been reduced to a watercourse, and has in places disappeared under roads and houses, daily becoming more difficult to find. Probably, on a normal summer weekend, the B&A now carries more boats than at any time in its existence.

REFERENCES:

1. Davies, Walter. *General View of the Agriculture & Domestic Economy of South Wales.* Vol. 2. 1814.
2. Hadfield, Charles. *Canals of South Wales & the Border.* D&C. 1967.
3. Hadfield op. cit.
4. PRO. MCC committee minutes, 16 October 1794.
5. NLOW. Maybery 428. The locations have been taken from a copy of Thomas Dadford's plan of 1792 in the writer's possession.
6. NLOW. Maybery 4031.
7. Figures taken from Priestley, J. *Historical Account of the Navigable Rivers, Canals & Railways of Great Britain.* London 1831.
8. Hadfield. op. cit.
9. At Maesgwartha, near Gilwem, established in 1606 by Richard Hanbury. See Rees, W. *Industry before the Industrial Revolution.* University of Wales Press, 1968.
10. NLOW. Maybery 370-4.
11. PRO. MCC committee minutes, 14 October 1794.
12. ibid. 22 September 1795.
13. PRO. B&A proprietors minutes, 15 October 1795.
14. ibid. 17 December 1795 and MCC committee minutes, 15 October 1798.
15. PRO. B&A committee minutes, 18 February 1796.
16. ibid. 13 February 1797.
17. Hadfield. op. cit.
18. PRO. B&A committee minutes, 4 December 1797.
19. Hadfield. op. cit.
20. NLOW. Maybery 383.
21. PRO. B&A committee minutes, 26 July 1821.
22. NLOW. Maybery 784.
23. ibid. 772.
24. PRO. B&A committee minutes, 15 September 1802.

25. NLOW. Maybery 803 and 852.
26. ibid. 801.
27. NLOW. Maybery. 802.
28. ibid. 803.
29. ibid. 807-813.
30. PRO. B&A proprietors minutes, 21 October 1802.
31. Hadfield, Charles. *Canals of the West Midlands.* D&C. 1966.
32. PRO. B&A proprietors minutes, 26 April 1804.
33. PRO. MCC committee minutes, 19 February 1794.
34. PRO. B&A proprietors minutes, 15 October 1807.
35. PRO. MCC proprietors minutes, 4 November 1807.
36. PRO. B&A committee minutes, 17 August 1809.
37. ibid. 4 August 1809.
38. ibid. 7 February 1812.
39. ibid. 30 May 1811.
40. ibid. 4 March 1817.
41. PRO. MCC committee minutes, 6 May 1817.
42. ibid. 28 August 1818.
43. Hadfield, Charles. *Canals of South Wales & the Border.* D&C. 1967.
44. PRO. B&A committee minutes, 21 October 1819.
45. ibid. 12 March 1812.
46. PRO. B&A proprietors minutes, 15 October 1812.
47. NLOW. Maybery 923.
48. ibid. 924.
49. *Cambrian,* 6 February 1813.
50. ibid. 13 March 1813.
51. NLOW. Maybery 935-6.
52. Brecknock Museum, Brecon. A/79/2/3(57).
53. PRO. MCC proprietors minutes, 7 May 1828.
54. PRO. MCC committee minutes, 18 February 1829.
55. ibid. 2 June 1830.
56. Brinore Tramroad Co. minute book, 27 October 1828.
57. PRO. B&A committe minutes, 21 April 1832.
58. PRO. MCC committee minutes, 6 September and 6 November 1833.
59. ibid. 5 November 1834.
60. PRO. MCC committee minutes, 23 June 1843 and 12 December 1843.
61. PRO. B&A committee minutes, 20 December 1836.
62. ibid. 4 September 1845.
63. ibid. 6 October 1845.
64. ibid. 29 January 1846.
65. ibid. 28 January 1847.
66. *Hereford Journal,* 5 November 1845.
67. PRO. B&A proprietors minutes, 26 April 1849.
68. Hadfield, Charles. *Canals of South Wales & the Border.* D&C. 1967., Article on Monmouthshire Canal.
69. PRO. MCC committee minutes, 23 July 1851.
70. PRO. MCC proprietors minutes, 19 March 1852.
71. PRO. B&A committee minutes, 3 April 1852.
72. PRO. MCC committee minutes, 16 April 1852.
73. PRO. B&A committee minutes, 29 July 1852.
74. ibid. 27 January 1853.
75. MacDermot, E. T. *History of the Great Western Railway.* 1936.
76. PRO. B&A proprietors minutes, 27 April 1854.
77. Barrie, D.S.M. *The Brecon & Merthyr Railway.* Oakwood Press, 1957.
78. Barrie. op. cit. and Clinker, C. R. *The Hay Railway.* D&C. 1960.
79. PRO. B&A committee minutes, 1 February, 6 March, and 26 April 1860.
80. ibid. 3 November 1864.
81. ibid. 18 October 1866.

Hill's Warehouse, Llanfoist.

3

Usk (or Mamhilad) Railway

Six months after the B&A failed in its attempt to secure the annulment of the eight mile clause in its Act of 1793, plans were deposited for a tramroad from the canal at Croes y Pant (SO.314 041), near Mamhilad, to the western end of Usk Bridge (SO.374 007), a distance of five miles three chains.[1] An Act for its construction received the Royal Assent on 16 June 1814 (54 George III. ci). It is not known why this procedure was adopted, in preference to a request to the B&A to construct it under the powers of their Act.

The capital was to be £6,000 in shares of £50 each, and a further £3,000 might be raised by subscription. The subscribers named in the Act included Sir Robert Salusbury, representing the coal interest, Thomas Hill and Samuel Hopkins, of Blaenavon, and Watkin George of Pontypool (partner of C. Hanbury Leigh), representing the ironmasters. Banking was represented by three members of the Jones family of Llanarth, who had banking houses in Abergavenny and Pontypool. The engineer was John Hodgkinson,[2] and his son, Edmund who was later to be given the contract to build the Aberbeeg–Nantyglo tramroad for the MCC, was amongst those mentioned.

On 20 October 1813, the B&A committee had made arrangements with Francis MacDonnel, the Usk solicitor, that cargoes received for shipment should be carried to their destination in the trams used on the tramroad, but should be charged only on the weight of the contents. It has been impossible to ascertain what goods were to be carried requiring this special treatment, unless it was intended to be the products of the well-known japan works in Usk.

There is no trace of the Act having been implemented, and fieldwork reveals no trace of construction being started.

REFERENCES:

1. GRO. PBR. 21. Deposited 30 September 1813. No Book of Reference.
2. Priestley, J. *Historical Account of the Navigable Rivers, Canals and Railways throughout Great Britain.* London, 1831.

4

Craig yr Hafod Tramroad

To the south of Abergavenny, the Blorenge Mountain stands as a marker between the agricultural area of north Gwent, and the industrial district surrounding Blaenavon. South-eastward from the Blorenge is Mynydd Garn Fawr, forming the northernmost hill of the ridge dividing the Usk and Afon Llwyd valleys. Some 300 feet below the summit of Garn Fawr is the quarry which was the objective of this tramroad, some 900 feet above the Brecknock & Abergavenny Canal, where it had its lower terminus. At the top the country is wild moorland, and at the canalside the country is as lush as any in Gwent.

The quarry at SO.273 101 probably originally supplied limestone to the very old kiln at SO.2729 1031, and a gully behind the kiln would indicate that outcrop coal was worked at SO.2724 1039 to obtain fuel for it.

With the advent of the canal, there was a general tendency for kilns, formerly all close to the quarries, to be moved nearer the canal for greater ease of distribution. It was not surprising, therefore, that the assembly of 26 April 1810 should receive a request from a Mr. Walter Osland, for a tramroad to connect his quarry with the canal at the point where it crossed Nant Ochram (SO. 2972 0913). The B&A refused to make it, and informed Osland that he might construct it himself if he so wished.

A year later, on 24 April 1811, the B&A was requested to let or sell a piece of land below the canal in Llanellen parish by the Cwm Llanellen Lime & Coal Co. They replied that, at the time, they were not inclined to part with any of their land. It is possible that this was connected with a request for a tramroad from the canal to Llanover Court Farm, which would have provided a good market for lime. No confirmation of this has been found however.

An indication that a tramroad had been constructed from the quarry is given in the B&A minutes of 26 June 1812, when the Cwm Llanellen Lime & Coal Co. asked the company to substitute a drawbridge for the existing bridge at Twyn Glais (SO.3007 0967), which was unsuited to carry over a tramroad built on the level of the canal bank, due to the hump of all the canal bridges. The committee could see no reason why any alteration should be made, and the Lime & Coal Co. made no further attempt to cross the canal. Probably kilns had by this time been constructed on the canal's west bank, and it was desired to extend the tramroad over the canal to aid distribution.

The B&A was very disgruntled that there should be a tramroad constructed under its Act, from which it received no traffic on the canal, particularly so, as the output of the kilns was

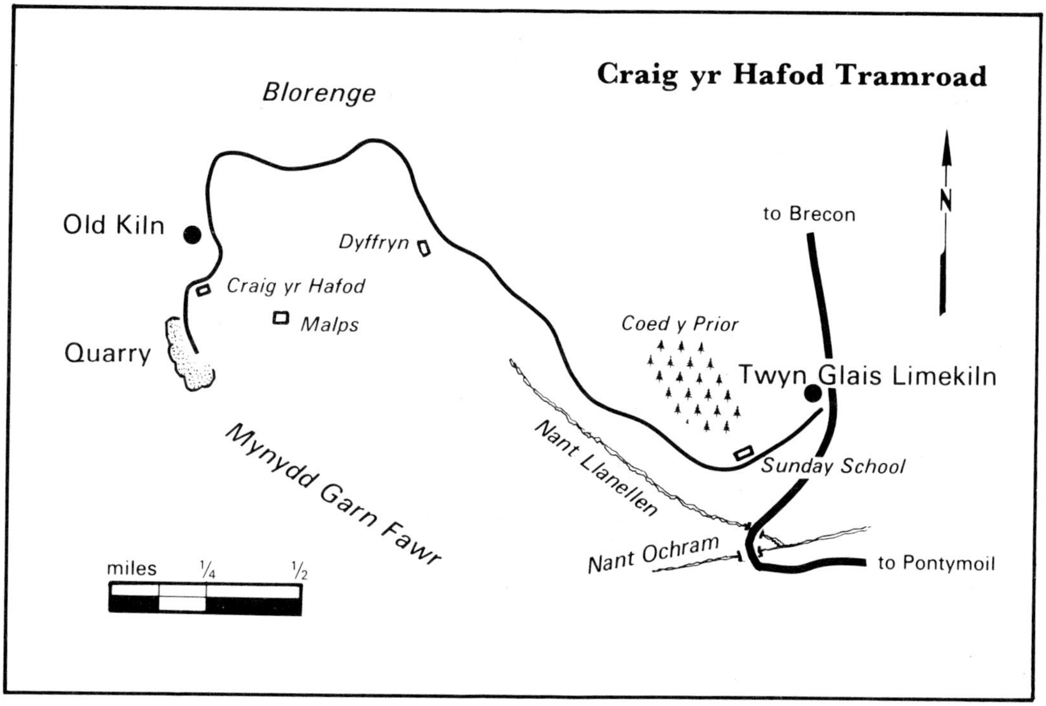

carried by road over a bridge they had been obliged to build, albeit to maintain a right of way that had existed before the canal. There were, at the time, two other schemes afoot to carry tramroads across the canal, one at Llanfoist and the other at Talybont on Usk, where George Overton planned to continue the Brinore Tramroad (Section 10) to connect with the Hay Railway.[1]

It was Osland, however, who was their main target, as in the other cases special bridges would have been necessary, for which B&A sanction would have been required, whereas Osland was making use of one that already existed. At the meeting on 18 October 1815, they instructed their solicitor to prepare a case for counsel's opinion, pointing out that the tramroad had been carried to the canalside to comply with the Act, but the company derived no benefit from it, not even the minimum tonnage of a quarter of a mile they could have demanded had the lime been ferried over the canal.

Counsel's opinion, dated 16 April 1816,[2] was that the proprietors of tramroads constructed under the Act were free to use as much or as little of such lines as they chose, and were under no obligation to carry any cargo on the canal; unless a bridge should be constructed specifically for their benefit, the B&A had no power to make any charge for crossing the canal; and finally that a landowner had no power to stop any cargo crossing his land on a legally constructed tramroad, and it would be pointless for them to try to get the landowners to do so. The B&A would have been in a far better position had they been willing to sell the land below the canal, and in a stronger one had they been willing to alter Twyn Glais Bridge to a drawbridge.

No information has been forthcoming with regard to the working of this tramroad, nor is it known when it ceased to work.

Route

Most of the earthworks of this tramroad are minimal, and difficult to find, and much is on agricultural land. Anyone seeking it should obtain permission from the occupiers.

It is evident that it was constructed as cheaply as possible, the blocks being much smaller than normal, and the fact that they show little wear, probably indicates that traffic was light, and possibly of short duration. Two culverts that have been found are of the simplest possible construction, being merely stone-lined channels with flat stone cappings. In its two and three-quarter mile course, it climbs approximately 900 feet, giving an average gradient of 1:16½, and in one place (from SO.2780 1084 to 2769 1084) it rises 100 feet in about 150 yards (1:4½).

There is no evidence that a wharf ever existed at Twyn Glais, nor of the tramroad having been carried to the canal bank. About forty yards from the west bank, and about twenty yards from the adjacent minor road, there now stands a shed used for agricultural purposes, the southern wall of which appears to have been the north-facing wall of the kilns containing the tapping mouths. A few yards to the south of this, an embankment, level with the top of the wall, marks the site of the tramroad (SO.3003 0966).

Continuing westward, the ground level rises above the height of the embankment and the tramroad is continued as a shallow cutting, running parallel with the road. The levels of the tramroad and the road coincide at SO.2985 0954, and here the tramroad crossed to the north of the road, and passing in front of Ty Porth (SO.2974 0953) it crossed the site of a recently built house, Coed y Prior Cottage, and keeping to the north of the road, can be seen as a level shelf about six feet above the road. Running to the rear of the erstwhile Sunday School (now a dwelling house) at SO.2957 0944, it crossed to the south of the road at SO.2945 0946, and round a spur over which the road is carried.

The levels of the road and the tramroad again coincide at SO.2924 0961, and the tramroad crossed to the north, at the entrance to a small quarry, and can be followed along the pathway through the wood (Coed y Prior). At the north-western corner of the wood, the line has been much disturbed by tree planting, but looking over the wall at SO.2894 0994, it is possible to see traces of the tramroad line leading away to cross the road at SO.2890 0998. The road at this point is some six feet below the level of the tramroad, but as this road is not shown on the 1″ O.S. (1st Edition) it is unlikely that any kind of bridge existed.

For the next quarter of a mile, the tramroad adopted a sinuous course along the side of Nant Llanellen to the west of the house, Bryn Deri, after making a hairpin bend in a small re-entrant at SO.2881 1010, where there are the remains of one of the rudimentary culverts. Keeping to a ledge above the stream, it proceeded by SO.2876 1007 and SO.2846 1016 (between which points many stone blocks are in evidence) to emerge on Coed y Prior Common at SO.2850 1032. Here it can be seen as a shallow cutting crossing the road to Ty Gwyn at SO.2836 1045.

From this point it continued on the line of the way-marked path leading to the Blorenge as far as SO.2824 1058, where the path to the Blorenge diverges to the north, and the tramroad continued its north-westerly direction to the gate at SO.2809 1075. Through the gate the tramroad took the right-hand fork of the path, through a small wood from SO.2806 1079 to SO.2798 1085, to turn westerly across the head of the valley, crossing the 1,000 foot contour at SO.2779 1090, and running west-south-west to SO.2736 1070, it crossed the second culvert.

To leave the re-entrant, the tramroad turned to run in a south-easterly direction for about 250 yards, to SO.2752 1054, whence it ran generally southerly close to the house, Craig yr Hafod (SO.2742 1025) running from SO.2746 1032 to SO.2745 1027 on a low, steeply inclined embankment of crude construction, apparently made by constructing two dry-stone walls, infilling with rubble, and using the blocks bearing the rails as the upper tier.

After crossing the track adjacent to Craig yr Hafod, the tramroad passed through the gate, to turn immediately to the west, and then to contour round to the south, to the quarry at SO.273 101.

Measurements taken at several places where blocks have been found opposite each other give an average distance of three feet three inches between the holes, indicating a gauge of approximately two feet nine inches, from back to back of the plates. An average distance between the blocks gives the length of the plates as three feet one and a half inches. A length of curved plate was found adjacent to Craig yr Hafod in 1962, and is now in Abergavenny Museum.

REFERENCES:
1. Brecknock Museum, Brecon. A.79/2/3(57).
2. NLOW. Maybery 973.

Hill's Tramroad: Cwm Llanwennarth.

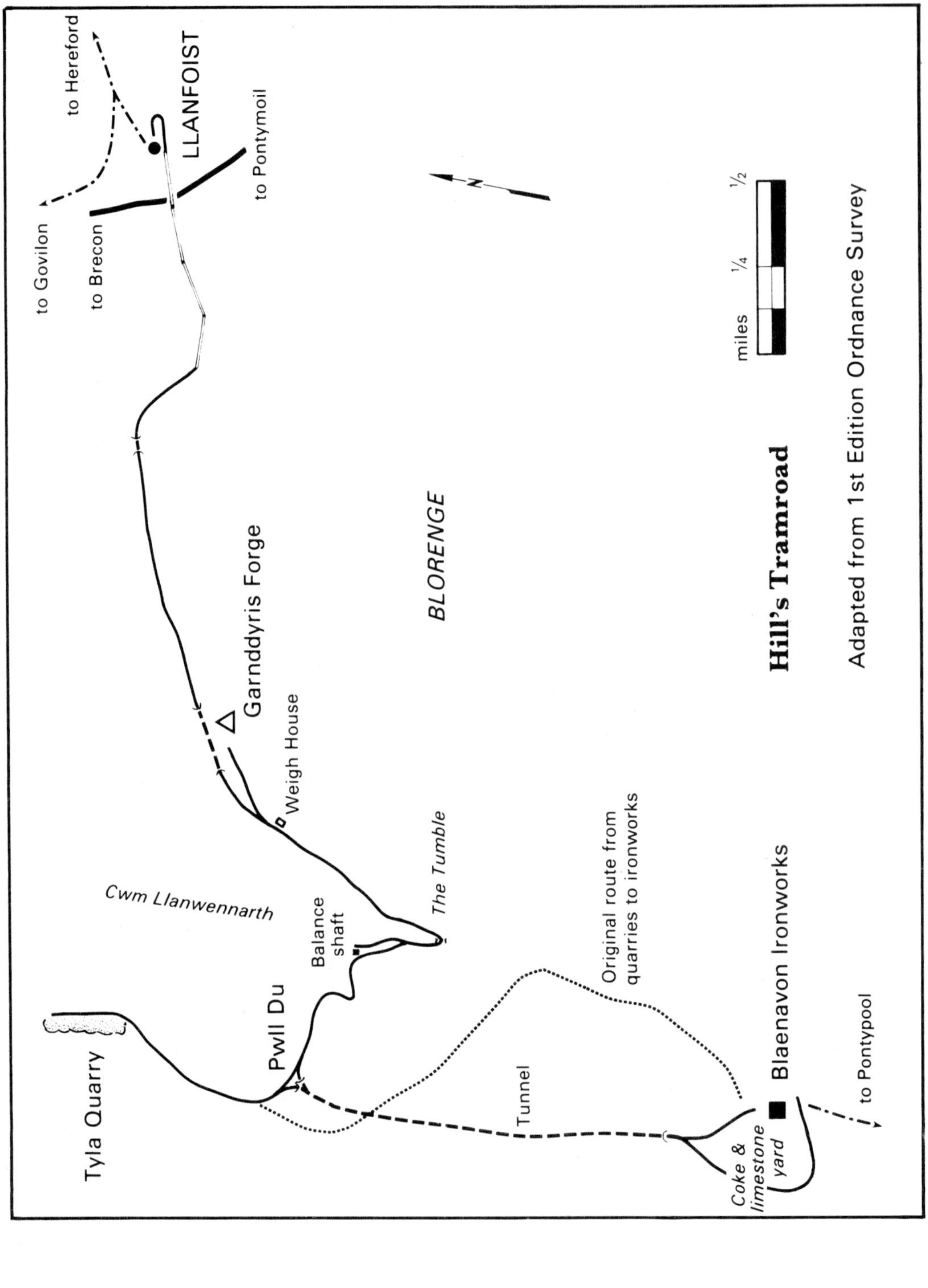

Hill's Tramroad

Adapted from 1st Edition Ordnance Survey

5

Hill's Tramroad

The Blaenavon Ironworks was established in 1789 on land leased from the Earl of Abergavenny by Thomas Hill of Dennis, Staffordshire and his partners Thomas Hopkins (his brother-in-law), and Benjamin Pratt of Wilden Ironworks, near Stourbridge.[1] Capital was obtained by a mortgage of Hill's Staffordshire estates to Bate & Robins, bankers of Stourbridge, a firm in which Hill was a partner.[2] The quantity of land obtained was too large for exploitation by the partnership, and large tracts were sub-let on which Nantyglo and Varteg Ironworks were established, there being sufficient mineral resources to satisfy the needs of all three.[3]

Thomas Hopkins died in 1791, leaving his share in the works to his son, Samuel; Pratt died in 1794, his share passing to Thomas Hill, the son of the founder, subject to trusts in favour of Pratt's wife and sister.[4] When Samuel Hopkins died in 1813, he bequeathed his quarter share of the works to Thomas Hill, junior.

Thomas Hill, senior, died in 1824 leaving his share of the works (and the debt of £121,000 due to Bate & Robins)[5] equally between his sons, Thomas and Waldron, who continued to run the works. Thomas, junior, died in 1827, leaving his eleven-sixteenths share to his son, also Thomas, who ran the works with his uncle until the latter's retirement from active participation in 1830.

The control of the Hill family ended in 1838, when a chartered company with a capital of £400,000 was formed to take over the business, under the name of the Blaenavon Iron & Coal Co., who traded with varying success until 1865; a limited company, the Blaenavon Co. Ltd. was then formed to take over the BI&CCo. Both the chartered company and the B.Co.Ltd. were bedevilled throughout their existences by the huge debt amassed in the early days of the works. This was not paid off until another company, the Blaenavon Iron & Steel Co. Ltd. took over the concern in 1870, and traded successfully until well into the present century.

By June 1793 the Blaenavon Works had been connected to Pontynewynydd by the Monmouthshire Canal Company's railroad,[7] and shortly after through transport to Newport was provided by that company's canal.

In 1794, the MCC was pressing the B&A to complete their canal to Pontymoil, and realising that as soon as this had been done it would be possible to tâke advantage of the cheaper rates of tonnage on the B&A, with those rates also applicable on the MCC line (under section 108 of the B&A Act), Hill, Hopkins & Co. applied to the B&A assembly on 30 April 1795, for a tramroad to be made from the intended line of the canal, near Llanfoist Church to a colliery that had been opened at Gwaith Newidd. It is possible that they also had ideas of entering the Breconshire coal trade.

The assembly gave their approval, but no action followed until June, when the committee received a reminder from Harrison (Hill's agent) and decided that a survey should be made by Hugh Henshall (for a short while the company's engineer), and three committee members. It was reported to the October assembly that the committee had not been satisfied by the engineer's report and had asked for guarantees (possibly in respect of the amount of traffic to be expected), which Harrison had refused to give. Hearing this, the assembly rescinded its previous decision to make the line, but gave permission for Hills to construct it themselves. Probably neither party realised that it would be seventeen years before it would be worth while.

The B&A was completed to its junction with the MCC in February 1812. In June 1814, Hill & Co. approached the B&A for a lease of a small piece of land at Govilon to construct a warehouse close to where the Blaenavon to Abergavenny turnpike crossed the canal. This was granted, but it was two years before the legal formalities were completed and the lease sealed. It is unlikely that it was intended to carry cargoes to Newport from this point in preference to using the direct route via Pontnewynydd, and it can only be assumed that distribution of goods by means of the turnpike was intended, or that it was hoped to establish a trade in the direction of Brecon.

In 1817, by which time the Blaenavon Works was controlled absolutely by the Hills, father and son, it was decided to diversify the business,[8] which hitherto had been entirely in cast iron, and a forge for the production of wrought iron goods was opened at Garnddyris, on the north of the Blorenge (SO.250 119).

From its inception the Blaenavon Works had used limestone from Tyla Quarries, and originally this was carried by road, and later by a tramroad running very much on the line of the present road from Pwll Du to Blaenavon; it is impossible to be more precise, as the area has largely been opencast, but it is shown on a plan dated 1819 as "Old Limestone Road".[9]

Archdeacon Coxe,[10] writing from observations made in 1800, describes one of the mine levels belonging to Blaenavon Works, as penetrating three-quarters of a mile into the hillside, and it was probably this level which was later carried through the mountain to Pwll Du (SO.245 117) and formed the basis of the tunnel through which a tramroad was carried, superceding the line over the top of the hill; it had been suggested that this took place in 1815.[11]

With the establishment of the Garnddyris Forge, it was natural that this route should form the basis of communication between the forge and the main works. The precise date on which this was completed cannot be determined, but possibly it was before the forge was opened, and it was certainly there by 1819, as shown by the plan previsously mentioned.

The Hills apparently still had the idea of a tramroad to the B&A in mind when they applied to the canal company for a plot of land adjacent to the canal at Llanfoist in August 1817, and in September two pieces of land, both on the line of their later tramroad, were purchased between Llanfoist and Garnddyris.[12] An application for a tramroad between these points was submitted to the assembly on 16 October 1817, and five members of the committee, any three of whom were to have the power to make decisions, were appointed to examine the proposed line. In particular, they were enjoined to examine the proposals for crossing the canal at Llanfoist. On 27 October, the sub-committee reported entirely in favour of the " . . . proposed Rail Way from Blaenafon Collieries to pass this canal and communicate with the Llanvihangel Railway . . . " The report was accepted, and a resolution passed that the iron company should be permitted to cross the canal "with Coal and other Articles making such compensation to this Company as shall be deemed equitable".

Not unnaturally, the Hills' idea of what constituted reasonable compensation differed considerably from what the B&A considered acceptable. Their offer to pay 1d. per ton for lime

Hill & Co's bridge over the canal at Llanfoist.

and limestone, and 2d. per ton for all other articles, including iron and coal, subject to a proviso that should the tonnage due for carriage of iron on the canal exceed the toll due for crossing it at Llanfoist, the latter should be waived, was entirely unacceptable, and the B&A replied with their proposals that Hills should pay 2d. per ton for lime and limestone, and 4d. per ton for all other goods, adding that it was solely their desire to accommodate them that led them to propose these terms.[13] At the assembly on the next day, Baker Gabb of Abergavenny, the Hills' solicitor, stated that he was not authorised to accept such terms by his principals, and the whole matter was passed to the committee.

On 6 May 1818, the committee received a letter from Thomas Hill, junior, accepting their invitation to meet them for discussions. The meeting took place on 11 May, and Hill insisted that it was essential that agreement should be reached, as until this was done he could not decide on the line for his incline. The outcome was the following agreement to take effect from the following 25 December:

1. Hill & Co. were to erect a bridge over the canal to the east of Llanfoist aqueduct to connect with their inclined plane "now forming," in such a manner as to avoid obstruction of the canal.
2. The agreement was to last for three years, with liberty for Hill & Co. to terminate it at six months notice, after the expiry of the first year.
3. Hill & Co. were to pay 1d. per ton on lime and limestone, and 2d. per ton on all other goods carried over the bridge.
4. Hill & Co's. books were to be open to inspection by the B&A, and should there be any dispute, the the canal company was to be at liberty to erect a weighing machine actually on the bridge.
5. The bridge was to be removed at the end of three years unless the agreement had been renewed in the meanwhile.[14]

Canal house and warehouse, Llanfoist Wharf. Note the 'winding hole' cut illegally by Hill & Co.

No firm date has been found for the completion of the tramroad, and deeds examined show that the necessary land was acquired at various times from September 1817 onwards. Those prior to 1820 refer to land on which it was proposed to construct a tramroad,[15] and those of later dates mention land on which a tramroad had been made.[16] There was considerable delay with part of the land required for the incline as the agreement with Frederick Secretan, Gabb's partner, was not completed until 10 December 1822.

Probably the line was operational by the end of 1821, Hill & Co. most likely obtaining permission to continue their works pending final settlement with Secretan. However, it is difficult to reconcile the fact that the 1819 plan shows against Pwll Du Quarries, "Limestone is worked here for Llanfoist Limekilns", with the fact that the first of the land leases was not completed until 24 June 1819, particularly as the plan suggests that full communication had already been established.

It was brought to the notice of the B&A committee in August 1820 that a presentment had been laid before the local magistrates that the road leading from Llanfoist up the Blorenge Mountain, which passed under the canal, was in need of repair. The parishioners had filed a plea of "Not Guilty", maintaining that the presentment was invalid as it did not refer to a road that the parish was obliged to keep in order.[17] The B&A was vitally interested in the matter, as should it be proven that the parish was liable, the road would be available for Hill & Co. to transport their goods, free of toll, from the canal to the transfer point with the Llanvihangel Railway in Llanfoist village. In March 1821, B. A. Griffiths, the canal Clerk, wrote to the company's solicitors in Brecon to know what steps should be taken to combat the claim, and informing them that the parishioners were being urged by Thomas Hill, senior, Baker Gabb, and the Rector of the parish to withdraw their plea, and to assume maintenance of the road.[18] The matter was finally settled when a declaration was obtained from an old man, then living in Brecon, who had in his time filled every office in Llanfoist parish, stating that the road in question had never to his knowledge been chargeable to the parish.[19] Doubtless, until the outcome of the affair was known, Hill & Co. delayed construction of the bridge over the canal!

Hill & Co. returned to the attack on the subject of charges for crossing the canal at the assembly of 25 April 1822, when they proposed that in the event of their carrying all the output of Garnddyris Works, and half of that of Blaenavon along the canal between midsummer 1822 and midsummer 1823, the tolls for crossing the bridge should be halved; and should the whole of the produce of the two works be carried on the canal, charges for crossing the bridge should be commuted to a flat rate of 10s. for the year. The B&A agreed, and in the light of later events it must be assumed that the arrangement was renewed from year to year.

In the intense economy drive instituted by the B&A in the closing months of 1822, the company took advantage of the fact that Llanfoist Wharf was used almost exclusively for Hill & Co's. traffic to appoint Hill's wharfinger there to issue "permits of lading" in respect of shipments by other parties. For this service he was to be paid the munificent sum of £5 a year, but in all probability he was adequately recompensed for the small additional work involved.[20]

With the death of Thomas Hill, senior, in 1824, followed in 1827 by that of his son, the Blaenavon Works devolved on Thomas Hill (grandson of the original Thomas) and his uncle, Waldron Hill. A new arrangement was made in respect of the Garnddyris Works by bringing in two new partners, Charles Wheeley and William Morgan, both described in the partnership deed as "Iron and Steel Manufacturers of Abergavenny".[21] The new firm, styled the "Garn Derris Iron Co.", was an extension of the existing partnership of Hill, Hill, Wheeley and Morgan, bankers, of Abergavenny.

No doubt, this move was inspired by the pressure being exerted by the Stourbridge bankers, Bate & Robins, for the repayment of the long outstanding loan made in 1789 to Thomas Hill, (the grandfather), to enable him to start the Blaenavon Works. In view of the large amount owing, approximately £121,000, the bankers insisted in 1830 that Thomas and Waldron Hill should mortgage their personal property, in addition to the mortgages they already held on the works.[22] The new partnership would enable the Hills to preserve a small part of their personal monies should the bankers foreclose.

A memorandum accompanying the mortgage of December 1830 specifically excluded the Garn Derris Iron Co. from the property charged, and in addition permitted that company to continue to work the Bodolog Coal Level, and the ponds that supplied water power to the works. It also recognised the practice of carrying Bodolog coal to Llanfoist on the tramroad, laying down that tonnages of 2d. per ton per mile should be paid. From the memorandum it is evident that the coal wharf and lime kilns at Llanfoist belonged to the Garn Derris Co. The tonnages were to be paid either to the mortgagees, of whom Thomas Hill was one in his capacity as a partner in Bate & Robins, or to Thomas and Waldron Hill as owners of the tramroad.

Relations between the B&A and the Hills seem to have been good after the agreement of 1822 until October 1835, when doubts were expressed in the assembly that Hill & Co. were abiding by the agreement, and the committee was ordered to investigate. Twelve months later, it was reported Hill & Co. were sending some of their produce via the MCC railroad and canal; James Peirce, now the B&A's Clerk, was instructed to ascertain the quantity involved, and to demand payment of the tonnages lost to the B&A.

Peirce reported on 16 January 1837 that he had found that Hill & Co. had sent 11,300 tons of iron on the MCC railroad, paying tonnages of £597 7s. 7d. and he had according to instructions demanded this sum of Thomas Hill, who had offered to pay £298 4s. in full settlement. Peirce was ordered to inform Hill that the committee would be willing to settle for £400, on condition that, in future, he should keep his agreement, and send all the output of the Blaenavon Works via the B&A. His reply was received on 27 April, enclosing £400 in respect of tonnages owed by "the late Blaenavon Company" up to Michaelmas 1836.

In June 1836, Thomas Hill who was in charge at Blaenavon, his uncle Waldron being then resident in Weymouth,[24] accepted an offer of £222,050 for the works from a consortium of London merchants,[25] and the Blaenavon Iron & Coal Co. was formed to take over the business, the capital being raised by an issue of 8,000 shares of £50 each. The formal Deed of Settlement was not executed until 1838, but the new company took over the works with effect from the beginning of 1837, the old partnership keeping it going until then. The BI&CCo. obtained a Royal Charter in April 1839.

The BI&CCo. had to honour all agreements entered into by the old company, of course, including that in respect of the canal bridge. It appears that the Garn Derris Iron Co. was also included in the deal, as right from the beginning their minute book refers to the Garnddyris Works as part of the BI&CCo.

It would have been expected that the opening of the Hereford Railway in September 1829, completing the tramroad route to that city from the Abergavenny district, would have given some impetus to the traffic carried on Hill's Tramroad. Ten years prior to the opening of the Hereford Railway, the *Hereford Journal* of 23 June 1819 reported that coal sales at Llangua, then the tramroad terminus, amounted to eighty tons daily (approximately 24,000 tons annually), which was insufficient to meet all demands. There are no figures available for the quantities carried on Hill's Tramroad whilst it was controlled by the family, and for only a few years (1837-43) is a figure mentioned in the BI&CCo's. minutes. These are not always quoted on the same basis; sometimes the actual tonnage is stated, and at other times the amount realised from sales; on several occasions only the profit on transactions is quoted. It can be deduced, however, that the average quantities involved were about 16,000 tons of coal and 6,000 tons of lime annually; quantities which on the 1819 figures would hardly have saturated the market!

The BI&CCo. was always under capitalised, and for the management the early years consisted of finding ways to improve liquidity and reduce the costs of their operations. With a view to the latter, Richard Johnson, their secretary, wrote to the B&A on 16 October 1844, suggesting that the toll for crossing the canal should be reduced to a nominal wayleave of 10s. a year, in view of the fact " . . . that no injury is done to the canal, either by impeding the passage of the boats or depriving it of a trade which could not, under any circumstances, pass along it . . . ".[27] The B&A replied that they could not conceive any reason for altering the arrangements made in April 1822.

A year later, Johnson returned to the attack, and was again referred to the existing agreement, and was sent, in addition, a bill for £125 0s. 11d. in respect of tolls incurred in the year that had passed.

Truth to tell, the years 1841-44 had been a very bad period for the BI&CCo. In 1843 it proved impossible to borrow more money from their bankers, and in spite of a large reduction in all grades of staff from miners to clerks, the directors found it necessary to inject their own money to keep the concern afloat.[28] By 1848, trade had taken an upturn, and the company was in a position to consider expansion of its mining activities and improvements in its manufacturing methods.

With a view to improving its mining operations and its transport system, Thomas Dyne Steel was appointed Assistant-Manager and Engineer, and manager of the 'Sale Coal Collieries'.

Steel, the son of the works surgeon, although only twenty-six years old, was an experienced civil engineer, and immediately set to work to improve the company's internal transport. Bearing in mind the imminent alteration of the MCC's Blaenavon Rail Road to a locomotive railway, he introduced standard gauge rails and steam haulage to a large part of the company's lines.[29]

The Pwll Du tunnel had always been a bottle-neck in the transport between the main works Garnddyris and the B&A canal. To obviate the difficulty, Steel constructed standard gauge inclines from Blaenavon and Pwll Du, with a winding engine at Incline Top (SO.238 111). The standard gauge waggons each accommodated four of the two foot gauge trams used on the tramroad. Probably the inclines were completed in 1850.

After a lapse of five years, the BI&CCo. returned to its attack on the B&A, and on 17 October 1850, the committee was informed that the iron company was determined to put an end to its present agreement and to cross the canal at Llanfoist without paying tolls, quoting their solicitor as the authority for stating that such tolls were *ultra vires*. The B&A solicitor was unable to find any flaw in the BI&CCo's. arguments, and, not knowing what to do, the committee allowed matters to drift until 16 October 1851, when a letter was received from Johnson, refusing to pay £106 19s. 6d. demanded for tolls to midsummer 1850. Discussion of the letter was put off to the next meeting.

They were stung into activity on 13 November 1851, however, when the solicitor informed them that he had been served with notice that the BI&CCo. was demanding repayment of all tolls paid in respect of the canal crossing over the last thirty years, and that they had applied to

Hill's Tramroad: Pwll Du Tunnel, north portal.

the court for a ruling that they might cross the canal free of payment and as of right. The solicitor was instructed to take counsel's opinion, and to call a special committee meeting when it was received.

Whilst awaiting the opinion, a sub-committee was formed to visit the bridge site to ascertain precisely the state of affairs there. They reported on 11 February 1852, that the bridge and part of the inclines on either side were constructed on canal property, and that the abutments of the bridge had been built on land obtained by filling in part of the canal, considerably reducing its width. In addition, a "winding-hole" for turning boats had been cut out of the off-side of the canal. The solicitor was instructed to pass this additional information to counsel, and as it improved the B&A position against the BI&CCo. he was instructed also to demand payment of the arrears of tolls, if necessary taking them to court.

Counsel's opinion was received on 20 May, together with his suggestions for a suitable letter to send to the BI&CCo. However, as the MCC was almost ready to open the Blaenavon to Pontnewynydd line as a railway, and the Newport, Abergavenny & Hereford Railway was already under active construction, it was evident that the B&A would have to be most circumspect in its approach to the iron company as these two railways would together form a good route for all products of the BI&CCo. Instead of sending the suggested letter, it was resolved that the BI&CCo. should be sued for the outstanding tolls up to September 1851, and that they

Hill's Tramroad: above Pwll Du Quarry showing Balance Pit.

might continue to use the bridge subject to the B&A receiving a full list of all goods passing since that date on which tolls should have been paid.

The action against the BI&CCo. was successful. As a result they agreed to pay £153 16s. 5d. in respect of arrears, and to enter into a new agreement, which was signed on 1 March 1853. Under it the iron company agreed to pay 1d. per ton on all goods passing over the canal, after paying in full all tolls up to 25 December 1852; to supply a certified copy of their accounts relating to traffic on the tramroad with all future payments, which were to be made quarterly, and if, in any one year, the whole produce of the works was carried on the canal, the toll should be satisfied by the payment of a nominal wayleave of 10s., the onus of proving this latter point was placed squarely on the BI&CCo. They were to pay a rent of 1s. annually for the use of the B&A's land that had been taken, and the arrangements were to be renewable annually.[30]

About 1847, the BI&CCo. commenced to move the site of its smelting operations from their original leasehold site, to a freehold one to the west of the valley, at Forgeside, where there was room for their propsed expansion, and in 1853 it was reported to the shareholders that it was intended to move the forge from Garnddyris to the new site, commenting that:

> "... The site of the mill at Garnddyris has been the cause of constant inconvenience and expence on account of its distance from Head Quarters at Blaenavon, and now that the Monmouthshire Canal Company have reduced their tonnages, and improved the mode of conveyance along their roads, these disadvantages are more strikingly felt; the transit from Garnddyris being by a circuitous and expensive route of Twenty Four miles, whilst the Port of Newport will be reached by a locomotive railway from Blaenavon of only Sixteen miles ..."[31]

The report was delivered to the shareholders on 22 April 1853, seven weeks after the conclusion of the new agreement with the B&A! Two months later it was announced that it was intended to erect a new forge and mill at Blaenavon to replace Garnddyris.

Progress on the new forge was slow, and whilst it was reported to have been started in the director's report of 21 December 1858, it was not until May 1862 that it was reported to be in use. In the interval, Garnddyris Forge would undoubtedly have been run down, but the actual date of abandonment is not shown in the BI&CCo's. minutes.

Once the run-down of Garnddyris was started it is unlikely that any iron was sent out via Llanfoist, and any traffic carried on the tramroad was probably in coal and limestone. It is unlikely that much of the coal tonnages mentioned in the minutes followed this route other than what was required for lime burning and local sale.

On 13 March 1860, a part of the canal bank above Llanfoist village collapsed, and the resultant rush of water did considerable damage, including washing away a portion of the lower incline from the canal to the kilns,[32] and it is probable that no attempt was made to re-open the incline; certainly no further limestone sales are recorded in the BI&CCo. minutes. This is confirmed by the B&A minutes of 17 August 1861, when the circumstances under which the BI&CCo. had removed the bridge and tramroad from the company's property were ordered to be investigated. It will be recalled the under the original agreement of 1818, the bridge was to be removed after three years should there not be a renewal, but there was no similar clause in the 1853 agreement. As the bridge is extant (1978) it can only be assumed that the B&A forced the BI&CCo. to replace it.

The 25″ O.S. of 1879–80 shows the tramroad to have been removed at Garnddyris, but the land on which it was laid remained in the possession of the Blaenavon Iron & Steel Co. Ltd., as successors to the BI&CCo., until 1891 when an Order of Exchange was obtained, and the Earl of Abergavenny resumed possession in exchange for other lands passed to the company.[33]

Whilst, in common with other tramroads constructed under the B&A Act of 1793, Hill's Tramroad must have been available to the public for use with their own vehicles, there is no evidence that use was made of this facility, other than the permission granted to the Garnddyris Co. in December 1830. However, its existence was certainly used as an inducement in the letting of at least one colliery; an advertisement in the *Hereford Journal* of 28 December 1836 stating that " . . . iron and coal are within an easy distance of the Brecknock & Abergavenny Canal near Llanellen to which they may be conveyed by a Tram-road and an incline plain . . . " It is unlikely that anyone but the Blaenavon company made use of it however.

There were two serious accidents on the incline in 1845, both involving persons illegally riding on the trams; in one, a girl insisted on riding on top of a loaded tram and fell off under the remainder of the train, and in the other, a boy attempted to jump from one tram to another whilst it was in motion, and the wheels of a following tram passed over his leg.[34]

Route

Whilst the remains of the blast furnaces on the site of the original Blaenavon Works are possibly the finest memorial extant to the era of stone built furnaces, subsequent industrial activity in the course of the last 100 years since they were last used, has successfully contrived so to alter the surrounding land that it is now virtually impossible to locate the routes of any of the tramroads serving it within half a mile of the works.

As far as the southern tunnel entrance, the route has been plotted as accurately as possible with the aid of Thomas Deakin's plan of 1819.[35] Similarly the line of the tunnel must be treated with reservation, as there are now no landmarks left on the surface, due to opencast operations.

Leaving the tapping side of the furnaces, the tramroad ran south-westerly from SO.249 092, to turn to run north-westerly for about 400 yards to SO.246 095 where it turned to run slightly east of north, crossing the Brynmawr–Blaenavon road (B4248) at SO.248 096. This wide detour was needed to avoid running through the coke-yard. Continuing in the same direction it entered the tunnel at SO.248 098. The mouth of the tunnel has now been bricked up, and a few yards to the west of it lie the sparse ruins of Tunnel Houses.

Deakin's plan shows that from the tunnel mouth there was another tramroad running south to the charging bank of the furnaces, crossing the B4248 at SO.249 096, where it joins the original tramroad over the hill from Tyla Quarries. Deakin labels the line from the tunnel to the charging bank as "the limestone road", which may well be evidence that the tunnel was in use to bring stone to Blaenavon before the establishment of Garnddyris Forge.

The tunnel runs straight, slightly east of north for about 300 yards, and then turns gradually to the north-west as far as SO.244 113. then bears to the north to emerge at Pwll Du, SO.2454 1168. Deakin's plan shows only one entrance to the tunnel at this point, but an unsigned plan of 1836[36] shows another entrance a few yards to the east, with a triangle of track the apex of which is underground. There is a further triangle shown outside the western entrance. The eastern entrance would seem to have post-dated the other.

From the tunnel, a line leads westward bearing to the north, but this is likely to have been a later line constructed after the introduction of Dyne Steel's inclines, to connect with the earlier quarry line over the hill.

Eastward from the tunnel, the tramroad passed to the south of Pwll Du village hall, SO.2463 1168 (now an adventure centre) and along the south of a wall to SO.2475 1156, where double walls indicate the point at which transfers were made to Steel's inclines. From here the

Hill's Tramroad: Cwm Llanwennarth looking towards Garnddyris.

tramroad can still be seen as a track leading around the rim of the quarry at SO.252 115, and passing the water-balance pit by which stone was brought up to the level of the tramroad at SO.2515 1146. Descending slightly more sharply, the track continues via SO.253 115 and SO.254 113 to cross to the eastern side of Cwm Llanwennarth at SO.2548 1123. The culvert which originally carried the tramroad over the little stream at this point is now merely a heap of stones.

Slightly to the west of the culvert, a line branched off to the west, leading to the quarries at SO.252 115; this can no longer be followed as it is blocked by a landslip at SO.254 114. Possibly it was this slip that led to the introduction of the balance pit, as whilst the 1819 plan shows this branch as "branch of road to the Limestone quarry" the 1836 plan shows it as a dotted line, presumably indicating that by that date it was out of use.

On the east of the Cwm, the tramroad runs north-north-east on a shelf cut into the hillside to reach the Garnddyris Works site at SO.258 119. At SO.2514 1181 can be seen the remains of the weigh-house shown on the 1819 plan. The tramroad passed to the west of the works at a lower level and was covered over from SO.257 119 to SO.258 120, to avoid scoria from the works falling onto it, the tunnel being made by walling and covering. The works was served by a line leaving the "main" close to the weigh-house.

From the works, the tramroad went north-easterly, to cross the former Blaenavon to Abergavenny turnpike road (now B4246) at SO.2599 1216, and contoured round the slopes of

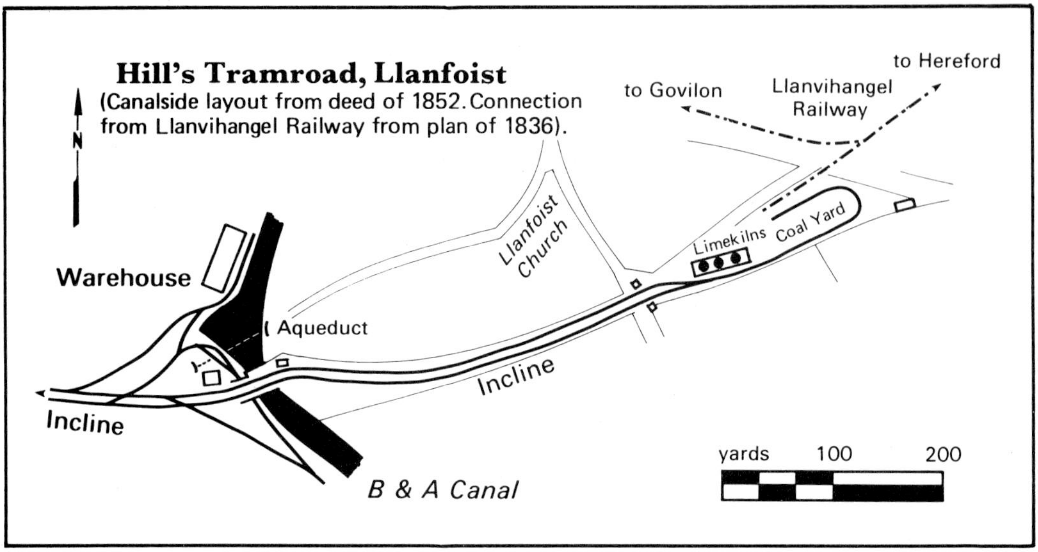

the Blorenge Mountain on a wide track to the head of the Llanfoist inclines, passing through a short tunnel (about forty yards) at SO.272 129.

The incline head was at SO.2775 1269, whence the first stage fell to SO.2799 1272, a steep fall of about 250 feet for 270 yards (about 1:3). The second stage was not as steep, falling 220 feet in some 310 yards to SO.2829 1288, and the third, again very steep, fell 300 feet in 370 yards to SO.2848 1297, where it was some six feet above the level of the B&A. The stages were all straight but not in line with one another, and between each there is still a small level platform, which must have accommodated a brake wheel, and at which trams must have been detached from one chain and fastened to another; this must have been an awkward process in the confined space of the platforms. There are now no traces of the brake wheels.

Canal House, on the canalside, is a three storied building, with small windows in the upper floor, so positioned that it was possible to look up the inclines from one end, and down the final stage from the other. The former warehouse at SO.2546 1304 has channels cut into the rock of its floor to the gauge of the tramroad, and probably at one time trams were run directly into it, although the plan accompanying the 1853 agreement does not show any connection. It is now in use as a boat repairing shop.

The continuation of the tramroad to the coal and lime yard in Llanfoist crossed the canal at SO.2850 1300, by a bridge composed of cast iron sheet, laid on cast iron "T" section girders, which is still in use as access to Canal House. The bridge was the head of the final incline, which had a fall of 120 feet in 290 yards to SO.2875 1314. Near its head, this incline bears evidence of the damage done in 1860, when the canal bank burst.

At the foot of this last incline the tramroad crossed the B4269 at SO.2875 1314, and thence to the top of the kilns at SO.2885 1325, it ran level on the top of an embankment. A steep, sharply curved extension led down from the kilns to the coal yard, on the level of the kiln mouths.

The Llanvihangel Railway ran through Llanfoist on the north side of B4246, and a short branch from SO.2885 1329 led into the coal yard to provide onward carriage to Abergavenny and further north.

Measurements taken at several places on the inclines at Llanfoist and in Cwm Llanwennarth show this plateway to have used a gauge of two feet, back to back of the plates, the rails being three feet in length. Through running with the Llanvihangel Railway was not possible, as that line employed a gauge of three feet six inches.

In 1973, bar-rails, probably dredged from the canal, were photographed lying on exposed blocks on the Llanfoist incline; this has since received some publicity. This type of rail was **not** used on Hill's Tramroad, and must have originated either on the Blaenavon to Pontnewynydd Railroad, the Trevil Rail Road or the Clydach Rail Road.

REFERENCES:

1. For general information in respect of the ironworks see Lloyd, J. *Early History of the Old South Wales Ironworks, 1760-1840*. 1906.
2. GRO. D.751/213-4.
3. GRO. D.751/328.
4. GRO. D.43/1800.
5. GRO. D.751/213-4.
6. GRO. D.751/327.
7. PRO. MCC proprietors minutes, 28 April 1794.
8. Rees, D. Morgan. *Mines, Mills & Furnaces*. H.M.S.O. 1969.
9. GRO. Man/A/2-273.
10. Coxe, W. *An Historical Tour of Monmouthshire*. London 1802.
11. *General History of the Garnddyrys Iron Works* in *Presenting Monmouthshire,* Journal of the Monmouthshire Local History Council, No. 37. Spring 1974.
12. GRO. D.751/140-1.
13. PRO. B&A committee minutes, 29 April 1818.
14. GRO. D.751/53.
15. GRO. D.751/140, 305 and 328.
16. GRO. D.751/71, 300, 325 and 328.
17. PRO. B&A committee minutes, 28 August 1820.
18. NLOW. Maybery 1005.
19. NLOW. Maybery 1006.
20. PRO. B&A committee minutes, 16 October 1822.
21. GRO. D.591/3.217.
22. GRO. D. 751/213-4.
23. PRO. B&A committee minutes, 20 October 1836.
24. GRO. D.751/213-4.
25. GRO. D.751/325.
26. GRO. D.751/356. Blaenavon Iron & Coal Co. proprietors minutes, 20 April 1839.
27. PRO. B&A committee minutes, 17 October 1844.
28. GRO. D.751/356, 28 April 1843.
29. GRO. ibid. various dates.
30. Original agreement held by Inland Waterways Board, Gloucester.
31. GRO. D.751/356, 22 April 1853.
32. *Cardiff & Merthyr Guardian,* 17 March 1860.
33. GRO. D.751/215.
34. *Hereford Journal,* 12 July, and 8 October 1845.
35. GRO. Man/A/2-273.
36. GRO. Man/A/2-279.

Llanfoist. Site of Hill's kilns and coal-yard. Llanvihangel Railway ran on the line of the double row of kerb stones, the siding going up the road alongside the inn.

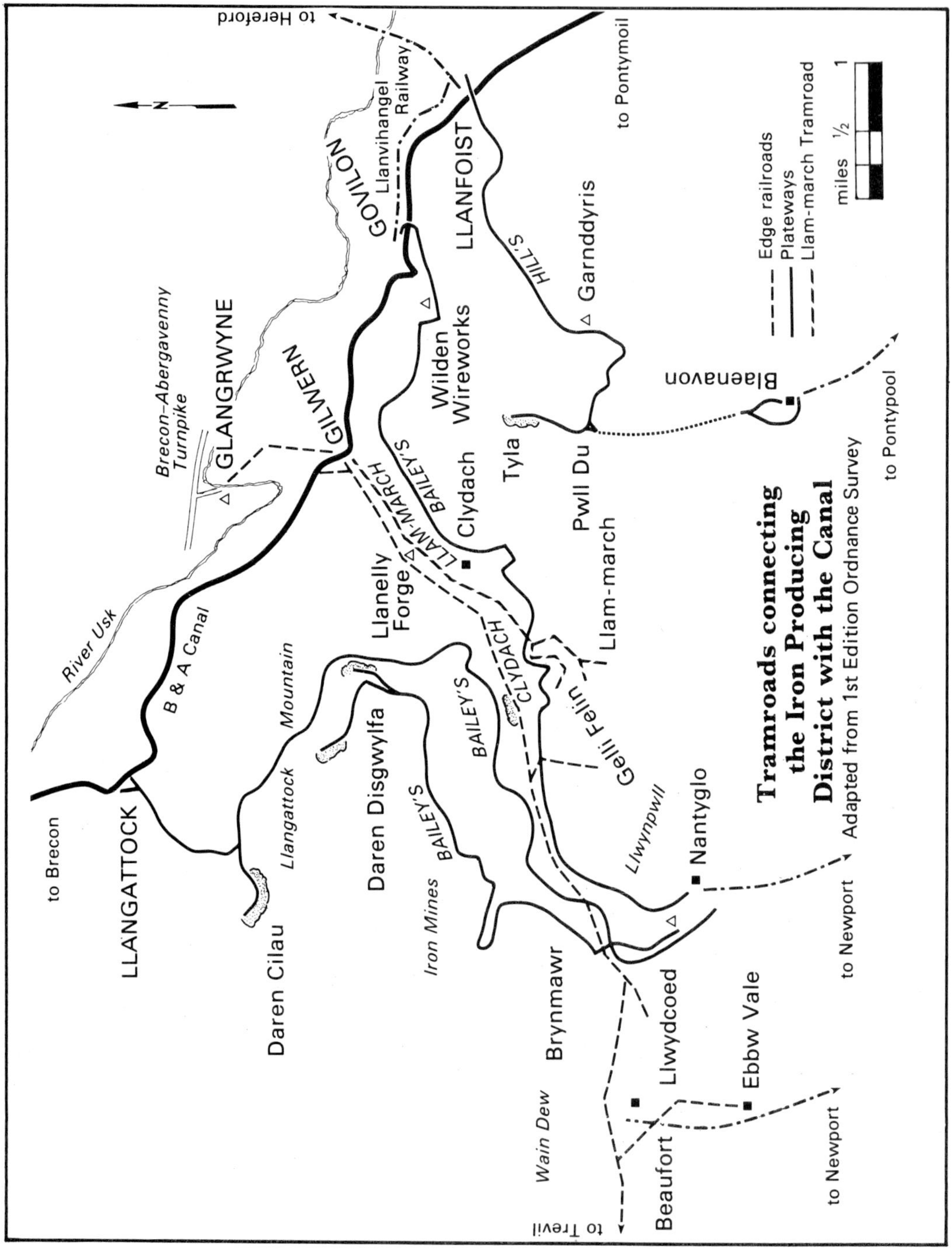

Tramroads connecting the Iron Producing District with the Canal

Adapted from 1st Edition Ordnance Survey

6

Bailey's Tramroads

There do not appear to be any surviving papers relating to the day to day working of the Nantyglo and Beaufort Ironworks, the only documents in respect of either that have been traced are those relating to land acquisition, which were of necessity passed to the Nantyglo & Blaina Ironworks Co. Ltd. when the two works were taken over by them. Much of the history of their rail communication has, therefore, been gleaned from the minute books of the B&A and the Monmouthshire Canals, and documents relating to other matters.

In addition to the edge-rail connection to the Clydach Rail Road which was, of course of the same gauge, there were three lines of plateway, two for the import of raw materials, each with its own gauge. It would be interesting to know what, if any, arrangements there were for interchange of loads between them.

The Nantyglo Ironworks had an exceedingly chequered early history. In 1791, Thomas Hill of Blaenavon realised that the land held by him and his two partners was larger than they could utilise profitably, and proposed to Harford, Partridge & Co. of Bristol that he and they should establish a works at Nantyglo. A formal deed of partnership was entered into in April 1792, and in 1794, R. S. Harford, acting on behalf of the Bristol partners, informed Hill that the works were ready to start production. Hill flatly refused to let the project continue, claiming that he had only undertaken to provide £10,000 capital, and that sum had been overspent by £485. Always an awkward man with whom to have dealings, he entirely ignored the fact that Harford, Partridge & Co. had, in fact, spent £16,381 of their own money. Finding himself unable to get Hill to change his mind, Harford discharged all the workmen except those he had engaged to tend the furnace, and put it into blast to use up the accumulated raw materials, to sell as much as possible in order to reduce the loss his company had incurred. Production ceased in 1796.

Although Harford continued to live in Nantyglo until 1799, when the Bristol company took over the Ebbw Vale Ironworks, there was no further activity at Nantyglo until 1802, when Hill formed a new partnership consisting of himself, his agent Joseph Harrison and a Blaenavon inn-keeper by the name of John Griffiths, calling themselves the "Nantyglo Company".[1] Lucrative contracts were obtained, but disagreements arose between the partners,[2] and in 1805, the MCC minuted that Harrison, who seems to have been in charge, should be approached for their overdue tonnages, and if there was no satisfactory answer, Hill should be brought into the dispute.[3] By 1808 matters came to a head, and the works was again idle.

After three years of idleness, the works was let to Joseph Bailey and Matthew Wayne on 25 March 1811 for the remainder of the term for which it was held by the Blaenavon partnership, at a rental of £2,050 a year, plus an annual charge of £750 in respect of every furnace erected above three in number.[4] Probably the capital for the new enterprise came from Joseph Bailey, who had recently benefitted under the will of his uncle, Richard Crawshay, of Cyfarthfa Ironworks, Merthyr Tydfil. Wayne, who had been employed at Cyfarthfa, undoubtedly supplied the technical know-how. Wayne left the partnership in 1820, to establish his own works at Gadlys, Aberdare, and his place at Nantyglo was taken by Bailey's brother, Crawshay Bailey. Under the Bailey brothers the works prospered, and in 1833 they purchased the Beaufort Works for £45,000, and thereafter the works were run as one concern, iron smelted at Beaufort being conveyed to Nantyglo to be processed at the large foundry they established there.

The Beaufort Ironworks was established by Edward and Jonathan Kendall on land leased to them in October 1779, by the Duke of Beaufort. This was another case where the quantity of land was too large for one works to handle economically, and a large tract was sub-let to the Clydach Ironworks of Frere, Cooke and Kendall, retaining land to the west of Brynmawr for their own use.[5] Iron production probably commenced at Beaufort in 1780–81, the iron being sent to Glangrwyne to be processed, until a forge was opened at the works in 1785.[6] Railroad communication was established between Beaufort and Crumlin in 1794,[7] and in 1799, the opening of the MCC canal provided through transport to Newport.[8]

In 1796, Joseph Latham was taken into the partnership, and after the death of Edward Kendall in 1807, and Jonathan in 1812, he ran the works with the executors of the original partners until Edward Kendall's son, also Edward, came of age in 1816. The original partnership was formally dissolved in April 1816[9] and a new partnership consisting of Edward Kendall, junior, W. H. Bevan and W. H. West was formed under the title of "Kendall & Bevan". The works continued in their ownership until taken over by the Baileys in 1833. Latham had two daughters, one of whom married Joseph Bailey, and the other W. H. Bevan, at the time Clerk and Treasurer of the Trevil Rail Road.[10]

Joseph Bailey died in March 1859, and Crawshay carried on the two works in the name of J. & C. Bailey, until 1871, when he sold the business to the Nantyglo & Blaina Ironworks Co. Ltd. The new company closed the Beaufort Works in November 1873, and in March 1874 the Nantyglo Works suffered the same fate, the production of iron being concentrated on Blaina. The further history of the new company falls entirely in the railway era, and is outside the scope of the present study.[11]

Nantyglo to Govilon

Possibly the greatest problem besetting the Nantyglo Ironworks in its early days was the difficulty in reaching the port of Newport. In the late eighteenth and early nineteenth centuries, most Welsh iron products were distributed either by coastal shipping or by boat up the Severn to the Midlands. Under its Act of 1792, the MCC was authorised to construct a railroad or a stone road to the works from its Beaufort Tramroad at Aberbeeg. Owing to the known unsettled state of the affairs of the Hill-Harford partnership, on 29 January 1794, the MCC turned down a request from them for the authorised railroad to be made. In retaliation, when the MCC ordered that all waggons should be accompanied by " . . . tickets . . . expressing the articles and weight it contains . . . "[12] in June of the following year, the iron company informed them that it would refuse to do so as long as the railroad remained unmade between Aberbeeg and Nantyglo. The MCC threatened that if they did not comply, they would indict them to the local magistrates for non-compliance with the by-laws.[13] The minute book does not record the outcome of the dispute, but in view of the imminent demise of the iron company it is unlikely that any action was taken.

Hill, Harford & Co. had no better luck when they tried to induce the B&A to lay rails on the existing stone road to the works from Brynmawr; on 4 June 1795, the canal Clerk was instructed to inform them that:

> " . . . the Committee does not think themselves justified at present to make the Rail Road now required by them . . . "

as it was generally understood that as soon as the existing stock of materials was used up it was intended to stop the works. He was instructed to add that as soon as there was a certainty that the works would be continued, or if the partnership would undertake to pay a rental of eight per cent of any sum expended on the line, the matter would be reconsidered.[14]

No request to either canal company was made by the Nantyglo Co., but in October 1810, Joseph Harrison asked the B&A for a railroad over a similar route to serve a colliery that he was opening slightly further down the Ebbw Valley than Nantyglo. It was turned down by the assembly, probably because they doubted his *bona fides*.[15]

A request by Bailey & Wayne shortly after they had taken over the works met with a very different reception. In response to a letter asking the B&A's permission to lay "plates" on the canal company's road between Nantyglo Works and "the upper Machine-House" on the Clydach Rail Road, the committee replied that there was no objection whatever, but that they would require an acknowledgement that the road was B&A property, and fixed £1 1s. a year as a suitable rent. Should Bailey & Wayne permit any traffic to pass which was not in connection with their works, the B&A reserved the right to take tonnages.[16]

This connecting line was completed by January 1814[17] and arrangements were made for free upward carriage of iron ore on both the canal and the railroad. The output of the works improved, and in 1817 it was estimated that 6,000 tons would be produced in the next year.[18] There is no mention in the B&A minutes of Bailey & Wayne making a formal agreement to send all their production along the canal, but it is probable that Nantyglo benefitted from the reduction of the canal tonnages on iron to 2d. per ton per mile negotiated by the Beaufort and Ebbw Vale Works in 1817. If so, it is probable that the arrangement for free upward carriage ceased to apply.

Possibly, it was this consideration that led, shortly after Crawshay Bailey's advent to the works in October 1820, to the application to the B&A for the construction of " . . . a Rail Way from the Canal at Llanwennarth [Govilon] to their iron Furnaces at Nantyglo . . . "[19] The line proposed was about a mile shorter than the Clydach Rail Road, and would reach the canal two miles nearer to Pontymoil, with correspondingly lower tonnages to be paid. However, Edward Frere, whose line from the Clydach Works to Llam-march was to be crossed at Craig Ffyddlwn (SO.2221 1248), lodged a written objection, and a decision was delayed until the assembly of the following April. In April 1821, the assembly refused to make the line, but gave tacit permission for the Bailey brothers to make it themselves, by permitting them to carry " . . . a Rail Way over this Canal near this Company's Wharf at Betty's Barn [SO.2727 1367] for the purpose of forming a Wharf and depositing of their Iron &c. and Coal to be conveyed along this Canal and the Llanvihangel Rail-way . . . " for which privilege a charge of 10s. a year was to be made.

In the absence of any direct evidence it is, perhaps, apposite to quote, more or less in full, the description of the opening of this plateway in the *Hereford Journal* of 19 December 1821, although this was in all probability culled from some other local newspaper, and, doubtless, somewhat embellished:

> "Abergavenny 10th. December: On Thursday the 6th instant, the only fine day we have had for a long time proved most propitious to the opening of the new tramroad from Nantyglo to Abergavenny, forming a junction with the Llanvihangel Road and forming a communication with a considerable part of Herefordshire, altogether a line of 24 miles. An idea of the energy with which the Road has been carried on may be formed from the circumstances of *it being completed in the short space of seven months,* regardless alike of the adamantine rocks and stupendous precipices impressing the most imposing consideration on the capital employed in this arduous undertaking . . . In the morning a line of trams containing 26 tons of coal with an immense lump weighing thirty-one hundredweight and forty tons of bars of iron moved off, preceded by a party in trams fitted up for the occasion, with a band of music. Approving cheers welcomed the procession along the route and were most heartily returned . . . At Abergavenny the bells rang all day and hundreds of inhabitants repaired to greet the procession. The whole of the coal was distributed to the poor, and will doubtless prove a source of great comfort in the cold weather we must shortly expect.
>
> A party joined a select number of respectable inhabitants at the Greyhound, where an excellent dinner was served up, after which the toasts were the King, the Church . . . the Lord Lieutenant, the Members for the County, Joseph Bailey Esquire, Crawshay Bailey Esquire, *the able engineer of the road, Thomas Hill Esquire of Blaenavon,* and many of the gentlemen present . . . We understand that a few gentlemen have it in contemplation to extend the road to Hereford . . . a reviving circumstance to the drooping spirit of the agricultural interest."*

One can well understand the rejoicing in view of the trade depression into which the country was drifting in the aftermath of the Napoleonic Wars.

A number of leases exist dated between March 1822 and January 1826, relating to the land required for the line in the parishes of Llanwennarth and Llanelly, and whilst there is no doubt that it was a plateway, it is interesting to note that those dated 1822–23 state that the land is required for the purpose of constucting a "railway",[20] those dated 1823 and subsequently, refer to "land now fenced and used as a railway".[21] In every case the plan accompanying the deed shows the line as a "tramroad". Plans drawn by David Davies of Crickhowell in 1824, showing the whole of the strip of land used for the tramroad, carry the caption "Survey and Plans of the Nantyglo Railway or Tram Road . . . "[22] thus providing another title for the line known locally as "Bailey's Tramroad".

In spite of the increasing prosperity of the Nantyglo Works, no plans were made by the MCC to construct the line from Aberbeeg to Nantyglo authorised in their Act of 1792, and no request apart from that of Hill, Harford & Co. in January 1794, seems to have emanated from the works proprietors prior to the construction of the line to Govilon. On 18 December 1823, seemingly off their own bat, the MCC committee resolved to construct a tramroad on the eastern side of the Ebbw Fach river from Aberbeeg to the Coalbrook Vale furnace of George Brewer & Co. and gave orders for a survey to be made. Wells, the MCC engineer, completed the plans in time to submit them to the meeting of 21 January 1824, but further consideration was deferred as no estimate of the cost had been made; it was decided, however, that should a later meeting decide to make the tramroad, it should be continued to the wall of Bailey's works. The February and March meetings again deferred making a decision and the matter was quietly forgotten.

Just over a year later, on 4 May 1825, the MCC assembly received three applications for tramroads in the Ebbw Fach valley, one from David Phillips, who wanted a line from Coalbrook Vale, one from William Brewer, on behalf of George Brewer & Co., and one from J. & C. Bailey

* The italics are mine. G.R.

who wanted a line made the whole of the way from Nantyglo. After special assemblies had been held on 15 June and 18 July the committee was given a free hand to get the tramroad made, and on 29 July, a resolution was passed, " . . . that Mr. Cooke do proceed to execute the line given notice of by Messrs. Joseph & Crawshay Bailey . . . "

Lack of funds caused considerable delay, and Edmund Hodgkinson, son of John, who contracted for the works, was several times told to work more slowly to make the money spin out,[23] but in January 1828, it was decided to " . . . urge him to complete the Aberbeeg Road . . . "[24] Young Hodgkinson appeared to be more interested in some sideline that he had found, more lucrative than constructing tramroads, and the contract was taken from him on 11 April.[25] A new contractor was found, and in July it was announced that the whole line of tramroad from Coalbrook Vale to Newport was " . . . nearly completed the Plates being all laid with the exception of Three-Hundred Yards at Crumlin . . . "[26] on the Crumlin-Risca line. In view of the resolution a couple of years later, by which the Baileys were to have the use free of tonnage of " . . . the part of the Aberbeeg Road built by them . . . ", stated to be about a mile,[27] it would appear that the MCC only constructed their tramroad as far as Coalbrook Vale, and that the Baileys had made the rest.

The Aberbeeg Tramroad was, naturally, constructed to the same gauge (four feet two inches back to back of the plates) as the remainder of the MCC tramroads, and it is an example of the far-sightedness of the Bailey brothers, that they had constructed their line from Nantyglo to Govilon to this gauge. They now had two good routes for their exports, on either of which the same stock could run, one of which could be used to restrain any excessive charge by the MCC.

Nothing has been found of the day to day working of the line to Govilon. In view of the steady gradient for almost the whole distance, it is considered quite feasible that trams were run down by gravity, as has been suggested to the writer by old persons living near Govilon. Another point which has been gleaned solely from local gossip is that at Siop Newydd (SO.2562 1376), where the shoeing forge for the tramroad horses was located, a dozen smiths were employed in two twelve hour shifts, in order to meet the demand. No documentary evidence has been found on either of these points however.

On 5 January 1855, it was announced in committee of the MCC that all tramroads had been converted to edge-railways, fit for locomotive working,[28] including the line from Coalbrook Vale to Nantyglo Market (adjacent to the Nantyglo Works), and it is probable that the advantage of transport to Newport by a means that did not involve transhipment, was shorter, and undoubtedly quicker than the tramroad-canal combination via the Govilon Tramroad had virtually caused that line to fall into disuse.

Plans deposited in 1858 for the Merthyr, Tredegar & Abergavenny Railway show the tramroad to be still in the possession of J. & C. Bailey and the railway as built and opened in September 1862, was constructed in some measure over the line of the tramroad. Crawshay Bailey who was now in sole command of the works, once again had matters work out to his advantage, as not only was he able to sell the line of the tramroad to the MTA, he was also made chairman of that company. Once the MTA reached Brynmawr, the line to Nantyglo Market was extended to sidings adjacent to the new railway in what is now Catholic Road., Brynmawr. Bailey again had two outlets for his produce.

Route

As this tramroad was not involved in obtaining raw materials its commencement in Nantyglo must have been on the tapping side of the furnaces, or in the foundry. This is borne out by David Davies' plan of 1824,[29] which shows the line to commence to the west of the, then, course of the Ebbw Fach river at approximately SO.190 110, slightly to the north of the forge site of later days, where it must be assumed that it connected with the internal lines of the works. Recent alterations, coupled with the building of the Brynmawr & Western Valleys Railway in 1905, and the culverting of the river whilst the Nantyglo Works was operative, have effectively combined to hide the site of its commencement.

From the works the tramroad ran in a straight line to SO.1929 1161, where it was later bridged by the MTA, and entered what is now Greenland Road at SO.1936 1161, and followed the line of the road to SO.1944 1193 where it turned to the east to join the line of the MTA at SO.1951 1204. With only slight deviations it then followed the line of the railway to SO.2082 1228 where it turned south-eastward for a few yards to run parallel with the line to Edward Kendall's Gelli Felen collieries from the Clydach Rail Road. The line to Gelli Felen was possibly disused by this time, and Bailey's line probably used the culvert over Nant Gam at SO.2085 1225 built for the other tramroad. This culvert is now very dilapidated.

Immediately south of the culvert, Bailey's Tramroad again turned to the east, keeping to a slightly higher level than the railway, which it crossed at SO.2125 1228 (the site of the later Gelli Felen Halt) and paralleled it, keeping to the north, as far as SO.2146 1214, adjacent to the western portal of Gelli Felen tunnel. Whilst the railway burrowed through the spur, the tramroad crossed the stream at SO.2149 1213, and passed along the street of the now deserted hamlet at SO.2157 1226. At SO.2164 1229 it turned south-east running on a ledge cut into the cliff face to cross the railway near the eastern end of the tunnel, and crossed the stream at SO 2184 1213 by a culvert which is in good condition, as the London & North Western Railway, which absorbed the MTA, put in flood control works in order to protect the culvert under the railway, and in so doing protected the tramroad culvert.

Turning north immediately after crossing the culvert, the tramroad crossed the line of the railway at SO.2181 1221, and ran parallel to it to cross again to the south at approximately SO.2206 1221, passing under the Llam-march Tramroad incline at about SO.2220 1247. All trace of it is then lost due to quarrying, but at SO.2257 1246, there are the remains of a culvert over the Sychnant which was adapted by the quarry-owners to their own purposes. To the east of the culvert, although much disturbed, it is still possible to trace the line of the tramroad on a higher level than the railway, which line it joined at SO.2177 1264, to cross and re-appear over the mouth of the northern of the twin Clydach tunnels at SO.2278 1264.

Here it entered the line of the existing track leading to Clydach village at SO.2308 1274, and it is necessary for the next quarter of a mile to rely on Davies' plan, as railway construction and quarrying have combined to conceal the line of the tramroad. At SO.2308 1274 it turned south to cross the Nant Dyar Fach at SO.2309 1264, adjacent to the culvert that carried the railway over the same stream. Turning northward, the tramroad came as far as the present road by the station, SO.2316 1269, then turned south-east to cross the Nant Dyar at SO.2326 1263, and again turning north rejoined the line of the railway at SO.2324 1298, passing over ground latterly used by the siding to the Clydach Quarries.

From this point the line of the MTA was again followed with very little variation, and Davies' plan shows roads in the positions where level crossings were adopted by the railway company at SO.2334 1330, SO.2356 1360 and SO.2439 1395 (Gilwern Halt). Continuing on the line of the

Bailey's Tramroad: Cwm Llanwennarth.

railway, past the quarry at SO.247 142 to SO.249 140, it diverged to the west of the railway. From here the tramroad has disappeared, but it can be picked up again as a low embankment at SO.2504 1384, running across the field to a gate at SO.2529 1370, and entering the lane past Pentre Farm (SO 2535 1375), to the crossroads at SO.2550 1365.

Falling a little it followed the track between stone walls passing Siop Newydd (SO 2567 1365), crossing the line of the railway at SO.2578 1366, and can be seen as vague traces in the field immediately to the east of the road bridge, curving to the south, to enter the railway cutting at SO.2395 1355, at the highest point of the cutting side. It left the cutting at SO.2597 1348, to run southward on the western side of Cwm Llanwennarth, turning by a hairpin bend over the stream at SO.2597 1335. The bridge over the stream is in a remarkably good state of preservation, but is suffering from the roots of a tree embedded in it.

Turning north-east from the bridge, the tramroad joined the railway at SO.2609 1341, at the eastern end of the railway viaduct passing Wilden Wire Works site. Between this point and SO.2620 1342, the tramroad has been completely destroyed by the railway works, and can next be picked up at SO.2623 1343, as a road. From here, the tramroad has been absorbed into the modern road system, via SO.2640 1346 and SO.2665 1356, from which point the road has, according to Davies' plan been made on the line of the tramroad to the canal bridge at SO.2717 1366, which the tramroad crossed, to turn immediately westward, descending to the level of the canal, running along the north bank to Bailey's warehouse at SO.2697 1369.

The deposited plans of the MTA show that there was a branch running from approximately SO.2615 1341 into the Wilden Wireworks at SO.2609 1348. After the construction of the railway a siding was run into the works in a straight line from SO.2625 1346.

Stone blocks found in Cwm Llanwennarth show that the track used on this part of the line was laid on cast iron tie-bars, with the ends of the ties supported on stone blocks. In other parts, evidence shows that it was laid with the more usual plate, spiked directly to the blocks.

Nantyglo to Daren Disgwylfa

Whilst the greater part of this two feet four inch gauge (back to back) plateway has been traced on the ground, documentary evidence is very sparse, and it is not known when it was constructed. It was in existence by 1818, when land was let by the Duke of Beaufort to Thomas Ellwood, the mineral agent of the Brecknock Boat Co., which is described as being "... bounded on the North by the Tram Road leading from Nanty Glo Works to the Disgwylfa Rocks . . . "[30] This has been identified from the plan accompanying the lease as from SO.1890 1198 to SO.1905 1208, in Clydach Street, Brynmawr.

A lease of 1821, by the Duke to Kendall & Bevan of Beaufort Ironworks, refers to one of the boundaries of the land concerned as " . . . on the West by a certain Fault which branches off in a Northerly direction from or near the Tunnel on the Lime Stone Road of Messrs. Joseph and Crawshay Bailey . . . "[31] This document is, fortunately, accompanied by a plan which has enabled the portions of the line that have been obliterated by mining to be located on the modern map.

Leases of other land in Clydach Street, Brynmawr,[32] refer to the tramroad as the "mine tramway". "Mine" was, at the time, the common term for iron-ore, which was obtained in the Clydach Terrace area (SO.190 132) through which the tramroad passed. These leases are of a later date, about 1870, by which time the works was using another source of limestone, and probably only iron-ore was being carried, the limestone traffic having been taken over by the tramroad to the Daren Cilau quarries in 1828.

The 25″ SO. (1st edition) 1880 shows the tramroad *in situ,* but the plan accompanying a deed of 1900 relating to the assignment of a lease in Clydach Street does not show the line. Remains that were excavated in the course of alterations to the street in 1976 show that the line was laid with cast iron plates laid in cast iron ties; partings had the plates and the ties cast in one assembly.

Route

The 25″ O.S. (1st edition) indicates that the tramroad to Daren Disgwylfa paralleled the connection from the Clydach Rail Road into the works, but no indication is given of the manner in which this importing line reached the furnaces. In fact, due to the construction of the Brynmawr & Western Valleys Railway, the whole of the site of it from the works to the line of the MTA has been lost, and the first place at which it can be identified is SO.1863 1147, where there is still (1978) the bridge by which it was crossed by the MTA in 1864.

The tramroad continued to run northward from the MTA, skirting the Welfare Park, formerly the site of one of the reservoirs supplying the Nantyglo Works. It turned eastward at SO.1859 1164 as far as SO.1868 1170 passing to the north of the triangle to cross George Street at

SO.1871 1174, and passing the length of Well Street, crossed King Street to enter Clydach Street at SO.1889 1189.

Running the length of Clydach Street via SO.1809 1201, SO.1904 1215 and SO.1901 1222, it left the site of the present housing estate at SO.1923 1244, and can be followed as the footpath turning west at SO.1917 1273, and continues as a ledge on the hillside running parallel to the Heads of the Valleys Road.

The 1821 plan shows the river Clydach to have been crossed at about SO.1862 1289, and the tramroad to have returned eastward to the north of the river as far as SO.1885 1289, where it turned northward. The whole of this section to the north of the river has been destroyed by the Heads of the Valleys Road, and the diversion of the river.

The tunnel mentioned in the 1821 lease appears to have been at SO. 189 120, and from this the tramroad ran south-east to SO. 1930 1281 where it turned to run in a generally north-easterly direction to SO.1954 1315, but the next point at which it can be fully authenticated is SO.2033 1332, whence it can be followed as a grass-grown track, on which there are occasional minor earthworks, to SO.2169 1380, where the track splits into two, the easterly running to Daren Disgwylfa Quarries at SO.219 144. In all probability this was the tramroad's original destination, and the westerly branch was a later addition.

The western branch crossed the brow of the hill at SO.2176 1437, immediately behind the Daren Disgwylfa Quarry, and turning north-west descended to Pant Gilwern, SO.2123 1460, then turning north, climbing to the quarries at SO.211 148. It is not known how far into these quarries this particular tramroad penetrated, but it is thought likely that the full extent of the quarries was not reached until after the tramroad to connect with the Llangattock Tramroad had been made.

Blocks at Pant Gilwern bear evidence of the use of cast iron ties, similar to those dug up in Clydach Street, Brynmawr.

Bailey's (Llangattock) Tramroad

On 24 April 1828, the assembly of the B&A received an application from J. & C. Bailey asking them to construct a tramroad from the canal near Llangattock to the Nantyglo Works. Consideration of the proposal was passed to the committee who decided on 24 July that it was not considered expedient for the B&A to make it, but that the Baileys might construct it for themselves. Despite the fact that Joseph Bailey was sitting with the committee, the Clerk was instructed to forward a copy of the resolution to the iron company.

No reason was given for the request, but it is probable that as the iron industry was still in the grip of the post Napoleonic depression, the Baileys were seeking an outlet for the produce of their collieries which could not be absorbed in the reduced iron making.

Little time was lost in commencing work, and in August 1828, Thomas Ellwood of the Brecknock Boat Co. wrote to his principals in Brecon informing them that:

> "... Mr. C. Bailey has been taking levels for the new Road through your Hafod Farm I don't know how it is between you and them concerning leave to go through the Farm but should the arrangements between you for them to pass (but which I don't doubt it will) I beg leave to make a few remarks which if they meet your approbation might be as well to draw up when you settle ..."

He went on to suggest that the valuable top soil should be spread onto the adjacent fields, that the tramroad should be walled to avoid sheep straying onto it, and that proper access should be provided for crossing the tramroad to reach fields otherwise cut off.[33]

It would appear that the Baileys had not consulted the BBCo. in respect of the trespass, but as soon as they realised that their scheme was known, they climbed down and wrote to Peirce, Lloyd & Co., the principals in the BBCo. in 17 September that:

> "We have conferred with Mr. Ellwood respecting our making the Road through your leasehold land at the Hafod and agree to the undermentioned conditions . . . "

and then detailed the points that Ellwood had raised, and concluded:

> "We feel obliged by your compliance and hope that this will be of mutual advantage to each of us".[34]

Matters were sufficiently advanced by October for the Baileys to approach the B&A for permission to make a wharf at Llangattock, which was granted.

In the following July, the BBCo., who were lessees of the Llangattock Tramroad (Section 9), notified the B&A that they has assigned their lease to J. & C. Bailey, whom it was stated intended to carry their tramroad to the head of the Llangattock incline, and the B&A were notified at the same meeting that the Baileys offered to purchase the incline at a price equal to the cost of construction.[35] No attempt at a valuation was made by either party until January 1831, when Ellwood was asked to prepare one. The assembly of 28 April approved the sale to the Baileys, but no price is mentioned in the minutes.

Probably the purchase coincided with the completion of the line from Nantyglo, and it is most likely that about the same time all the limestone traffic was put onto the new line, the Daren Disgwylfa Tramroad then only carrying iron-ore. Possibly the latter line may have been abandoned in part, as it is not shown on a plan drawn by Edward Davies in 1833 to show the communications between the quarries and the Nantyglo and Beaufort ironworks.[37]

The Beaufort Works was purchased by the Baileys in January 1833 and possession taken in March, although the deal was not completed until some time later due to the inability of Kendall & Bevan to produce all the title deeds. In August, the Trevil Rail Road Co. applied for and obtained an injunction to prevent the Baileys extending their line from Llangattock to serve Beaufort Works, claiming that under the original partnership deed, the TRR had been given the right to supply and transport all the limestone used at Beaufort. The Kendalls had been a party to this deed, and had undertaken that such should by the case for their *"heirs, successors and assigns"*. According to the injunction, the Baileys had been informed of this when they agreed to purchase the works, and Crawshay Bailey is reported to have said, "Well! We must manage as well as we can with the Trevil Rail Road and try to make some arrangement with them about it".[39]

It was not in the nature of the Baileys to accept such a setback, and on appeal the injunction was withdrawn, but was brought before the court again in November 1834, when no decision was reached. The TRR filed a Bill of Revivor in April 1835, and the case was only partly heard by July 1836, when the solicitors of the contending parties reached a compromise, under which the Baileys were to pay the TRR £2,125 as compensation for loss of tonnages, and the TRR was to withdraw the Bill and to release them from any further claim under the original Deed of Settlement.[40]

No date for the completion of the line to the Beaufort Works has been found, but it is fairly certain that it was fininshed well before the second court action, as in January 1834, the Clerk to the B&A reported that he had found only one place at which the canal company's railroad had been crossed without permission, which was at the head of the incline down to Beaufort Works; the committee decided to charge the Baileys a nominal rent for the crossing. It was the complex of railroads and tramroads at this point that forced the MTA, some years later, to construct Beaufort tunnel, where otherwise there could have been an open cutting.

The BBCo's. lease of the Llangattock Tramroad expired in September 1836, and in October the B&A accepted Bailey's offer of £105 annually for the tramroad, but the quarries at Daren Cilau, and the kilns at Llangattock were to remain let to the BBCo. It was probably at this time that a branch of the tramroad was put in to serve Bailey's coal wharf (SO.2075 1721), by a drawbridge over the canal.

Writing in 1844, Samuel Lewis in his *Topographical Dictionary of Wales*, under the heading 'Llangattock', states:

> "... There is also a tram-road and inclined plane for conveying limestone from the Darren Rocks to the canal; and Messrs. Bailey have established a steam-carriage and tram-road for the carriage of coal from their mines in Monmouthshire to the wharf near the village of Llangattock ..."

The mention of the "steam-carriage and tram-road" refers to the introduction of locomotive haulage on the line, probably about 1840. In view of the careful grading of the line, it is considered probable that the use of locomotives was planned from its inception; it rises only thirty-three feet in the first 1,400 yards from the works, followed by a rise of five feet in the next 600 yards. The remaining four miles seven chains, despite some slight undulations, are, in the main level.

Unfortunately, little seems to be known about the locomotives, other than that a later one was called "Cymro" (Welshman), which possibly formed the basis of the popular Welsh song *Cosher Baileys Engine*, of which there are many versions current, but very few in print! There is a public-house in Bailey Street, Brynmawr (which parallels the tramroad) called *The Cymro*, which possibly commemorates this locomotive.

The opening of the MTA had little repercussion as far as this tramroad of the Bailey's was concerned, the railway being carried over it where the two lines crossed to the west of Brynmawr station, and they ran parallel from Brynmawr to the top of Beaufort Hill.

When the B&A was sold to the MRCC in 1865, the Baileys purchased all of their lines to the west of Clydach Bridge, Brynmawr for £300.[41] The BBCo. ceased to trade about the same time, and that company's lease of Llwynypwll Colliery reverted to the Baileys as successors to the Kendalls. The B&A's Clydach Rail Road, running on the lower level in Alma Street, Brynmawr, was altered to run on the higher level to connect the line to Llwynypwll with the Llangattock line at the end of Lower Bailey Street to take Llwynypwll coal direct to Nantyglo. Probably, this section was altered to a plateway at the same time.

Route

The 25″SO (1st edition) shows two sites for Nantyglo Furnaces. One at SO.193 104 is marked "Old Furnaces", and the other is at SO.191 104. As Bailey's (Langattock) Tramroad was built to import raw materials it would have been directed to the charging-bank of the furnaces, and the general direction of the line makes it most likely that, as constructed, it served the "Old Furnaces".

It entered Limestone Road at SO.1933 1083 to run along it to SO.1913 1115, where it passed onto the embankment now known as Barley Field Road. Formerly, this embankment continued to the line of the MTA at SO.1910 1152, where it was bridged by the railway to the west of the station. Crossing The Square at SO.1909 1164, it entered Lower Bailey Street at SO.1916 1178, after passing *The Cymro* public house, some twenty-five yards away at SO.1912 1176.

King Street was crossed at SO.1924 1236, and it entered Intermediate Road. Bearing a little to the east at SO.1944 1236 it would have crossed the Heads of the Valleys Road at SO.1954 1236.

From this point the route is very easily traced, as it ran on the line of the tarred road leading round Llangattock Mountain via SO.1996 1237, adjacent to Rhydw Farm; SO.2056 1283, Hafod Farm; SO.2109 1279, Penylan Fach; SO.2200 1305, Pantmawr Quarry, and at SO.2209 1313, it threw off a branch to the quarry at Craig y Gaer (SO.223 132).

Now running northward, it passed the Daren Disgwylfa Quarries at SO.2208 1443 at a great deal lower level, and continued to SO.2206 1471, where it bore to the west and then north-west to SO.2161 1491, where there was a branch to Chwar Pant y Rhiw (SO.204 155). It is not known if this branch preceded the incline from SO.2043 1547 to the main line at SO.2064 1554, or if it was a later addition.

The tarred road finished at SO.2048 1571, but the tramroad continued as the track now much rutted by the parking of cars. Fortunately cars have only been able to traverse about 100 yards of the route, and the tramroad can be followed as a rough track to SO.2004 1591, where it meets the head of the inclines of the Llangattock Tramroad.

The line to the Beaufort Works branched from the Llangattock line at the southern end of Lower Bailey Street, Brynmawr, and ran straight, slightly south of west to SO.1884 1156, in Warwick Road to the south of the present Welfare Park, formerly a reservoir serving the Nantyglo Works. It passed the full length of Warwick Road to SO.1812 1149, crossing the line from Daren Disgwylfa at SO.1863 1146, adjacent to the MTA bridge over that line.

From the end of Warwick Road, the tramroad passed onto a long curving embankment, bearing slightly to the north via SO.1778 1159, where it curves to the south-west to resume its course parallel to the MTA at SO.1756 1150. At SO.1741 1148, it crossed over the top of the MTA's Beaufort tunnel, and (on the level) the Clydach Rail Road of the B&A; this latter road was virtually abandoned by 1833. The tramroad reached the head of the incline to Beaufort Works at SO.1734 1147, and descended by one stage to the works at SO.1705 1126.

Evidence on Llangattock Mountain shows that this line was a plateway to a gauge of three feet six inches (back to back of the plates), the same as the Llangattock Tramroad, and the plates were laid in cast iron ties. This is the only part of the line which has not been disturbed or tarred over.

REFERENCES:

1. Lloyd, J. *Early History of the Old South Wales Iron Works, 1760–1840.* 1906.
2. NLOW. Maybery 3814.
3. PRO. MCC committee minutes, 28 February 1805.
4. GRO. D.751/328.
5. GRO. D.591/42.9.
6. Gray-Jones, A. *A History of Ebbw Vale.* 1970.
7. PRO. MCC proprietors minutes, 18 April 1794. Report of Thomas Dadford, junior.
8. Hadfield, Charles. *Canals of South Wales & the Border.* D&C, 1967.
9. *Hereford Journal,* 4 February 1818.
10. Jones, T. *History of Brecknockshire.* Glanusk Edn. 1909. Vol. 1. p. 139.
11. GRO. *Summary of the History of the Nantyglo & Blaina Ironworks Co. Ltd.* by W. H. Baker, County Archivist.
12. PRO. MCC committee minutes, 23 June 1795.
13. ibid. 21 July 1795.
14. PRO. B&A committee minutes, 4 June 1795.
15. PRO. B&A proprietors minutes, 18 October 1810.
16. PRO. B&A committee minutes, 30 May 1811.
17. ibid. 17 January 1814.
18. ibid. 4 March 1817.
19. PRO. B&A proprietors minutes, 19 October 1817.
20. GRO. D.591/39.29., 39.92., 16.24., and 12.41.
21. GRO. D.591/39.21 and 39.26.
22. GRO. Man/A/2-278.
23. PRO. MCC committee minutes, 3 March 1826.
24. ibid. 10 January 1828.
25. ibid. 11 April 1828.
26. ibid. 11 July 1828.
27. ibid. 3 February 1830.
28. ibid. 5 January 1855.
29. See note 22 above.
30. GRO. D.591/7.233.
31. GRO. D.397/1672.
32. NLOW. Badminton papers, various.
33. NLOW. Maybery 3615.
34. Brecknock Museum, Brecon. Maybery Collection, Vol. 2, fo. 139.
35. PRO. B&A committee minutes, 23 July 1829.
36. GRO. D.591/46.118.
37. NLOW. Map Room.
38. GRO. D.397/1679.
39. GRO. D.43/1335.
40. GRO. D.591/39.1.
41. PRO. B&A proprietors minutes, 18 October 1866.
42. 1st. Edn. O.S. 25″, 1878.

7

Clydach Rail Road

The valley cut by the little river Clydach running north-eastward for three and a half miles from Brynmawr to Gilwern today gives little indication of the important role it once played in the industrial history of the district; only around Brynmawr are industrial scars to be seen. Yet, as early as 1606, Richard Hanbury established an iron furnace and a forge at the lower end of the valley, attracted to the spot by the abundance of water-power, and wood for charcoal, and the easy access to iron ore and limestone in the adjacent hills.[1] And as late as 1875 iron was still being made at Clydach Ironworks, about a mile upstream from Gilwern. Slightly away from these activities was the Glangrwyne Forge, on the north bank of the river Usk, established about 1720 to work iron carried there by pack animals from the somewhat primitive furnaces then existing at Ebbw Vale.[2]

In early days, when production was small, the advantages of the valley outweighed the disadvantages that later became apparent, and carriage by animals easily coped with the output, but with the larger production made possible with the introduction of coke-smelting, transport became one of the major problems, together with the disposal of the scoria in the narrow valley.

The possibility of coal being brought down the valley from the easily worked outcrops at its head was undoubtedly one of the considerations leading to the plan to construct a canal in the Usk valley by means of which it was planned to provide easy carriage to Newbridge on Usk, to which point sea-going vessels could penetrate at the time, 1790. Most certainly it was the main reason for the entry of the gentry of Breconshire into the plan which was then expanded to take in Brecon. With the entry of the Monmouthshire Canal Company into the scheme, and the subsequent alteration of the southern terminus to Pontymoil, on that canal, the Breconshire party insisted that they would withdraw their support should coal from the Clydach valley not be made a primary consideration.

Bowing to their demands the first assembly of the company held on 16 May 1793 ordered that " . . . the Rail Road from the Collieries to Langroiney (Glangrwyne) be first made . . . " and the committee, meeting on 14 June, instructed that the first part of the authorised railroad to be made should be from Glangrwyne to Cwm Gelli Felen " . . . opposite the Gelli Felen Coal Pits . . . " which Edward Kendall, of Beaufort Ironworks, was intending to open at SO.214 118. The committee also decided to obtain tenders for the bridge over the Usk at Glangrwyne, and engaged John Dadford to take charge of the works at a fee of £210, out of which he was to pay all assistants he might require for surveying the ground and preparing the plans.

The rails were ordered to be obtained from Samuel Homfray, then at the Penydarren Ironworks, Merthyr Tydfil, to be made to Dadford's specification, and "... on the same terms in respect of price and quality as rails are contracted for by the Proprietors of the Monmouthshire Canal . . . " Iron sleepers, should Dadford decide to use them, were to be obtained from the same source, and payment was to be by bills of exchange payable at one month; settlement to be made at the end of each quarter commencing from the date of first delivery.

Dadford was ready with his plans in time for the meeting of 27 June, and the committee agreed that the railroad should be carried from the Usk at Glangrwyne along the east of the Clydach to Aberclydach (SO.246 152) and to cross to the west there and so to the intended line of the canal at Gilwern, and into the Clydach valley keeping to the west of the river, passing near to Hanbury's furnace, to terminate opposite Cwm Gelli Felen (approximately SO.204 123). Dadford also produced plans for the bridge over the Usk, which were approved, and it was decided to advertise in the Gloucester, Hereford and Bristol papers for tenders for its construction. The same meeting authorised the acceptance of Homfray's tender of £7 per ton for rails, and 4s. 8d. each for sleepers. All reference to the MCC was deleted from the order and John Dadford was given a free hand to use his own design of rail. At the beginning of July, a letter from the Duke of Beaufort's solicitor in London stated that he could see no reason why there should be any objection by His Grace, and authorised the commencement of work whenever the committee should think fit.[3]

Philip Williams, of Crickhowell was appointed "Pay-Clerk" (otherwise Clerk of the Works) for the construction of the railroad and the bridges on 27 July 1793, and it was agreed to accept the tenders of William Prichard, a Cardiff builder, for the construction of the bridges over the Usk and Clydach. The contract for the Usk bridge was signed on 6 August by Prichard and his two guarantors, Wyndham Lewis of Llanishen, near Cardiff, and George Whatley of Cardiff. It provided that the bridge should be completed by December 1793, and maintained for seven years after completion, any necessary repairs to be executed within one month of a defect being notified. Included in the contract price of £900, Prichard was to find all materials except the six cast iron sockets required in Dadford's design to be sunken in the bed of the river.[4] The plan accompanying the agreement shows that the bridge was to be carried on vertical wooden piles placed in the sockets, with no form of cross-bracing or triangulation; no means of retaining the piles in the sockets is shown in the plan. An invoice of 22 January 1794 shows that the sockets weighed over three tons each.[5]

The agreement for the smaller bridge over the Clydach was signed on 28 September, John Dadford acting on behalf of the committee. It provided for a normal stone bridge, with "parapats" of "common walling", to be completed by the end of December, at a cost of £210; Prichard was, again, to find all materials and his profit within the contract price.[6]

A short length of embankment to the south of the river Usk was let under contract to Jonathan Gee. The contract price for this is not known, but the important point was that the plans provided by Dadford made no provision for culverts to relieve the pressure of water in the event of the river flooding; an omission that was destined to have serious consequences.[7]

The actual construction of the railroad appears to have been commenced on 3 July 1793, as a letter written that day by Walter Watkins to John Powell of Brecon states that "... I can now inform you that about fifty hands began to work this morning and as many more are engaged for next week and I hope over five hundred men employed in a fortnight . . . "[8] Watkins added that he thought the gauge to which the railroad was being built was too narrow, and should be increased by six inches, and that he considered the order to Homfray for "slippers" should be countermanded immediately. His argument was that the additional width would permit waggons carrying four tons to be used with very little increase in the cost of haulage and only a slight increase in the cost of construction.

Homfray also wrote to Powell four days later saying that he had heard of Watkins' ideas from P. Williams, and pointing out that " . . . another inconvenience will attend in that if you or the Monmouthshire Co. wish to travel from one Rail Road to the other you cannot do it on account of the Roads being of difft. widths . . . "[9] He continued that owing to a miscalculation, he had underestimated the cost of the sleepers, and that he would, naturally, take care that any further estimate should be for the correct amount. Neither letter seems to have been brought before the committee, so it must be assumed that Powell did not consider it worth pursuing the idea.

With the rails being made at Penydarren, carriage over the ten miles to Cwm Gelli Felen was a difficult matter, as the only turnpike road, at that time, from Merthyr Tydfil to Abergavenny led over hills to Brecon and then along the Usk valley. Consequently the contract for carrying went to a farmer who undertook to carry 100 tons at £1 1s. per ton. In view of the terrain, and the lack of roads he was not overpaid.[10]

As early as September 1792 doubts had been expressed by the MCC committee as to the suitability of iron sleepers,[11] and Thomas Dadford had been ordered to report on the advisability of using stone blocks *in lieu*. It may be coincidence that the B&A ordered iron sleepers from Homfray in June 1793 to John Dadford's specification, and in July the MCC were offered a new type of sleeper.[12] By October, the MCC ordered that iron sleepers should be "entirely rejected", and in the same month the B&A requested Homfray to desist from making any more until "the utility thereof has been better ascertained". Unfortunately, no example of either the B&A or the MCC pattern has been traced, and the design and the means of attaching the rail must remain a matter of conjecture.

By early October 1793, matters were sufficiently advanced for the committee to order Williams to continue the railroad further up the valley, as far as the already constructed stone-road to the Nantyglo Ironworks, and to the Llwydcoed collieries, the works to be terminated at a suitable place for the continuation to the authorised end of the line at Wain Dew. In view of his increased reponsibility, Williams' salary was increased by £10 10s. a year.

Homfray commenced delivery of rails and sleepers at the beginning of July 1793, and by the end of September, 7,559 rails and 5,396 sleepers to a total value of £3,221 6s. 10d. had been delivered.[13] The contract called for payments to be made three months after delivery by bills at one month date,[14] and on 19 November, Homfray wrote expressing surprise that he had not yet been paid the £3,000 due in respect of the earlier deliveries. He also stated that he had ceased to make sleepers but he understood " . . . the difficulty is overcome in laying them on the Monmouthshire Road and no complaints are now made and the Waggons travel very well upon them . . . "[15]

Delivery of the contract quantity of rails and sleepers was completed by the end of December 1793, and on 20 January 1794 Homfray delivered his final account showing that all told he had made 600 tons of rails and 6,792 sleepers, at a cost of £5,784 16s., and in addition had supplied three and a half tons of bar iron, for which he charged £59 10s. Over and above the contract, he had made some forty-six tons of extra rails, for which he charged the B&A, making a total indebtedness of £6,167 3s. 6d. against which he had already received £4,000. Written at the foot of this account there is a note in another hand dated 22 January 1794 recording the fact that nine rails and forty-eight sleepers had been received broken, and that Homfray had undertaken to replace them.[16] On that date the B&A committee had resolved that "the sleepers be not covered until viewed by three or more of the committee or some person deputed by them for that purpose".

As no further money was forthcoming, on 1 February Homfray again wrote to the B&A informing them that he intended to refuse to permit any further withdrawal of rails from Penydarren, and that it was useless for the B&A to suggest that the rails were faulty, as they had been examined by Williams. He expressed his willingness to have the stock inspected by the company, and should any then be found wanting they would be replaced or credited at the company's option.[17]

Probably £1,600 was paid about this time, and on 11 February Homfray wrote to John Powell, the Brecon solicitor, and Clerk to the B&A, informing him that in response to a letter Powell had written, he had permitted Williams to take away some rails, but that he would not be answerable for the quality. It is a pity that Powell's letter is not available for inspection, as it evidently contained plausible excuses for the non-payment of the account, to draw from Homfray the comment " . . . I believe no person entertains a higher respect for the professional Gentn. than I do but if there were fewer of them engaged in the business the Wheels would be less clogged . . . " Homfray also complained that in forwarding a bill at two months for £200 in further payment he had been sent a letter stating that the money was sent "to accommodate" him as no more could be spared, which provoked the comment " . . . this is language I am not accustomed to as a tradesman . . . " He concluded, that he proposed to attend the committee meeting on 13 March,[18] doubtless intending to have his say in person. The minute book does not record his presence at the meeting, but a statement of account, dated 12 March re-states all deliveries, and shows £5,600 deducted as received by trade bills, leaving a balance of £567 3s. 6d. due. From this a pencil note deducts £200, probably representing the payment alluded to in a letter of 11 February.[19]

For all Homfray's assurances on the subject of iron sleepers, the committee decided to have as little to do with them as possible and on 13 March 1794 concluded agreements with Griffiths & Sons and Maund & Son, both of Brecon, that each should supply 1,000 wooden sleepers by the following 15 April, to be followed by a further 1,000 each by 1 May, all at 1s. 6d. each.

The same meeting ordered a contract to be prepared with Walter Watkins " . . . to set out and make a sufficient wharf, warehouse and buildings at or near Langroiney for carrying on the business of the Canal of Rail Road . . . ", and further ordered the Agent to make a road from the north end of the Usk bridge to the village of Glangrwyne (SO.240 163) to provide a communication with the Brecon to Abergavenny turnpike road (now A40). The B&A was preparing to enter trade.

On 17 October 1793, the B&A received the first two of many requests through the years, for tramroads to connect with the as yet, unstarted canal. The first was for a line from Gelli Felen, where Edward Kendall was proposing to open collieries, and the second from Frere, Cooke & Kendall, who required one to serve the ironworks they were then building at Clydach. The latter eventually became the Llam-march Rail Road, and the former scheme was amended by an agreement of 28 May 1794 to a line from Gelli Felen (SO.214 118) to join the company's line at Pont Harry Isaac (SO.204 123), a point now covered by the Heads of the Valleys Road (A456 T). Under the agreement, Kendall undertook to carry all coal obtained at Gelli Felen on the Clydach Rail Road, and not to construct any railroad lower down the valley unless it was to the Clydach Ironworks. In return, the B&A undertook to construct the railroad across the valley, and to charge for coal carried to the canal as though the distance was three and a half miles, although this was some thirty chains short of the actual distance and, under their Act, they would have been entitled to charge four miles. Kendall was also to receive preferential rates for coal carried up the valley to his limekilns, and in the case of coal intended for "country" sale, he might carry on horseback on the line of the railroad.[20]

Clydach Rail Road: Bar-rails and spike.

The exact date on which the Clydach Rail Road was opened for traffic does not seem to have been recorded, but there are several pointers indicating that progress was made. On 22 January 1794, Walter Watkins was instructed to provide two "Rail Waggons" for the use of the company, and on 24 April it was sufficiently advanced to necessitate the passing of a by-law stipulating that the maximum load to be carried in any waggon should be three tons. On 28 May, Dadford signed the certificate that the Usk bridge was completed,[21] and on 16 October a William Powell was allowed £5 5s. compensation in respect of his mare which had been killed ". . . in consequence of a defect in the Rail Road". Althogether the picture is one of piecemeal opening, but with insufficient evidence to say which parts were opened by any specific date. At a meeting in May 1794 a further 100 tons of rails were ordered from Homfray, and probably by this time the extension to the Llwydcoed Collieries, authorised in October 1793, was well under way, as orders were given for an extension beyond Wain Dew to join the Trevil Rail Road, then under construction, at Tyr Meddig (SO.1720 1104). Probably because they were afraid that a junction at this point would give the B&A too easy an access to Ebbw Vale Ironworks, the TRR refused to permit the junction, but stated that they were prepared to allow one to be made at Rhydyblew.[22] The committee fell in with this suggestion in July.

Whilst negotiations with the Duke of Beaufort for the use of his land proceeded smoothly, difficulty was experienced with a certain Mrs. Mary Parry, who owned a small farm, part of which was on the line of the railroad between Gilwern and Glangrwyne. She refused all offers made for the land and demanded large compensation, finally prosecuting Dadford and Gee at Hereford Assizes for trespass. She claimed that the most valuable part of the farm would be cut off from the farmhouse by the railroad, but according to the depositions by both Dadford and Gee, the line that the B&A proposed to use was already a cart-track, and would not cut off any appreciable part of the farm, and in any case, the Act granted her the right to cross the line for any farm purpose. The case was heard in October 1794, and judgement given at the canal company's figure for the land.[23]

Walter Watkins was ordered to make two waggons in January 1794, and in September the company was invoiced with two tons fifteen hundredweight of wheels and brasses by E. & J. Kendall of Beaufort Ironworks.[24] In the books of the Brecknock Boat Co., a similar quantity is shown to be sufficient for twenty-four wheels, i.e. six waggons,[25] so it is reasonable to suppose that shortly after this date there were eight waggons available for use. By October it was considered that sufficient had been made and it was decided to dispense with the services of the carpenters, and the smith and his striker as soon as they " . . . have put all the Waggons and Barrows in repair as it appears there will be no further occasion for their services".[26]

Little is known of the construction of these waggons. The carrying capacity must have been about three tons, and under the company's regulations, each had to be fitted with "double brakes".[27] By a decision of 12 March 1795, it was ordained that all wheels should measure four and a half inches "within the crest", presumably meaning the width of the tread. As the length of the rail used was four feet, the waggons would have had a wheel-base in excess of this, to avoid excessive weight being put onto any one rail. On 6 September 1794, the committee ruled that any person hiring a waggon should undertake to keep it, and deliver it up in good order, and on 15 October it was decided that no waggons should be let out for a shorter period than six months, or at a hiring charge of less than 3s. per week. On the following day the assembly ordered that the Clerk was to report monthly on the number of waggons let out "distinguishing them by their particular numbers".

An unexplained part of the construction of the line is revealed by an invoice presented by Evan Evans on 3 October 1794, covering "2 miles 586 yards, being flat paving stones to lay under the sleepers . . . £34 4s. 4d." At the foot of the invoice, written in another hand, appears "1st November 1794. Mr. P. Williams will report in what parts of the Road this work has been done, and whether charged at a reasonable price, and actually necessary to the service of the Company".[28] Apparently it was also inexplicable at the time!

In January 1794, Philip Williams was appointed General Agent and Pay Clerk to the company at a salary of £140 a year, until the canal should be completed. In the following October, in addition to his duties of supervising railroad construction, he was put in charge of the collection of accounts and tonnages in respect of the railroad. His instructions were to collect the money once a month, and to pay it to the treasurer "within the week next after". By February 1795 it was found that he owed the company £323 16s. 9¾d. and was instantly dismissed and ordered to repay the money in full by 1 March, under pain of prosecution. He was succeeded by John Knowles, who was appointed on 12 February at a salary of £80 a year.

Of greater importance to the company was the collapse of the Usk bridge in a flood on the night of 10 February 1795,[29] and two days later the committee ordered John Powell to send a messenger to William Prichard and his two sureties calling on them to rebuild it, and to inform Prichard that he would be expected to attend their meeting on 19 February. By 29 April nothing had been done, and it was resolved to seek counsel's opinion as to Prichard's liability to rebuild, and in the meanwhile Knowles was to secure what remained of the materials.

Counsel's opinion given in the middle of May was that it was considered that the B&A should have a good case; Prichard's contention that the specific disaster which had caused the bridge to collapse was not mentioned in the contract was thought to be irrelevant, as no particular form of failure was specified. It was suggested that the best way for the company to proceed would be for them to rebuild the bridge and then to sue Prichard and his sureties for the cost.[30] A sub-committee was appointed to look after the matter, with instructions to advertise for a surveyor to superintend the rebuilding.[31]

On 4 June, three members of the sub-committee were detailed to get in touch with Wyndham Lewis, one of Prichard's sureties, to make arrangements for the rebuilding, and the committee undertook to back them in any arrangements they might make. They reported to the committee, on 29 June, that Lewis had refused to meet them, and it was decided to resort to the Courts to settle things. This, however, was more easily said than done, and it was reported to the August meeting that proceedings had been delayed until the next Assizes, due to the solicitor's inability to contact George Whatley, the other surety. The case was eventually heard at Hereford in the early part of 1796, and went entirely against the B&A, largely on the question of contributory negligence as revealed by the evidence of Jonathan Gee that he had been under contract to make a solid embankment leading to the bridge, with no provision for culverts to permit flood water to escape.[32] This embankment suffered severely in October 1795, as is revealed in Knowles' letter to John Powell of 17th in which he reports that on 15th there had been " . . . a most extraordinary Flood in Clydach far surpassing in magnitude the Flood in February last . . . I am sorry to inform you that it has damaged our Road pretty considerably . . . " He thought it would cost £70 to £80 to repair, but was thankful that the damage had not been more severe.[33]

The loss of the bridge seems to have caused some difficulty to Fothergill & Co., of Sirhowy Ironworks, in getting iron to the forge at Glangrwyne, and on 19 November 1795, they were given permission "to use the Boat lately built at Langroiney on account of this Company" (presumably in connection with the bridge works), provided that they undertook to "convey such coal and lime as shall be required for the consumption of the Country at the rate of 3d. per ton"; Walter Watkins, whose warehouse would be supplied with the goods, was to be under an obligation to maintain the boat, which should be at the disposal of the B&A and its workmen if required. A year later, on 20 October 1796, the assembly resolved to take the ferry into the company's hands provided that it did not involve an outlay of more than £100, with the idea of carrying the railroad waggons over the river. The committee reversed the decision on 12 January 1797, but nonetheless decided to leave down the rails to the weighing machine at Glangrwyne, which appears to have been on the north bank of the Usk.

Considerable inconvenience must also have been experienced by Walter Watkins in trading on the north of the river, as his supplies would, of necessity, have had to cross what Lloyd[34] described as a dangerous ford. A letter from him was read at the assembly of 27 April 1797 demanding that the bridge be rebuilt; the matter was referred to the committee, but appears to have been quietly forgotten. On several occasions in the course of the next twenty years the question of rebuilding was raised in committee, but on each occasion it was shelved. At length, in October 1855, a plan was put forward to obtain an Act to construct a bridge on the site by public subscription. The committee decided to oppose this, as it was intended to build on land which was still B&A property, but let to the Clydach Iron Co. Three years later, in July 1858, the B&A was approached by the Llangrwyney Bridge Co. Ltd. for permission to become joint tenants with the iron company in the road from the bridge site to Gilwern, and in the site itself. The committee agreed subject to the concurrence of the Clydach Iron Co., but it was not until 30 July 1863 that legal matters were sorted out, due to complications caused by the failure of the iron company in 1861, and the land and the road, from which the rails had long since been removed, let to the bridge company at £4 4s. a year.

II

On 24 July 1794, Edward Kendall entered into an agreement with the B&A that he should dispose of the whole output of his colliery at Gelli Felen for country sale at a reasonable price, as soon as the railroad to the colliery was ready. Presumably the railroad was completed by May 1795, as on the 20th John Dadford submitted his account in respect of his work on the Clydach Rail Road, and included the sum of £80 "To laying out and superintending . . . the Rail Road from opposite Cwm Gelli Felen to Fossalog . . . ".[35]

In August 1795, Kendall undertook to supply coal at 3s. per long ton[36] at the colliery, and the B&A ordered three waggons to be set aside to carry it from Gelli Felen to " . . . the Llanelly end of Llangroiney Bridge, there to be sold to the country on account of this Concern at 6s. 6d. per ton . . . " The bridge, of course, no longer existed, but the B&A was in business.

By October, Kendall had increased his price to 3s. 6d. per long ton and undertook to hold this price until the following April, and at the April 1796 assembly it was resolved that, with a view to promoting the coal trade not more than five miles tonnage should be charged on coal notwithstanding that it might have been carried above that distance. This did not, of course, affect Kendall whose colliery at Gelli Felen was the subject of a separate agreement limiting the tonnages to three and a half miles.

Kendall further increased his price in June to 3s. 8d. per ton to which the B&A had, perforce, to agree, as he stated that he could not afford to let them have it for less; he did agree, however, that he would hold it at this price for twelve months.

The committee realised that with Kendall as their only supplier they were being held over a barrel, and he could increase his price whenever he chose. In August they decided to find other

Gilwern: Clydach Wharf, terminus of the Clydach Rail Road. The feeder from the river is on the left. The wharfinger's house has now been demolished.

supplies, and instructed John Powell either to purchase or rent a colliery on the company's behalf on the best terms he could, provided that in the former case it should not cost more than £1,200, or if the latter, the rent should not be more than £100 a year for a term of twenty-one years. He reported in October that his search had failed, and it was decided to leave it to the joint Clerks, John and William Powell to treat with any party they might find who would be willing to supply up to 200 tons of coal weekly, for a period of twenty-one years, either at the colliery or at the canal. They were to be permitted to offer the inducement of a railroad of not more than a mile in length for the use of the supplier. The minute book does not record the result of the Powells' search, but by early 1798, the B&A seems to have handed over the coal trade to a consortium of canal proprietors consisting of the Clerk, John Powell, the banker, Jeffreys Wilkins, John Peirce and John Lloyd, all of Brecon, who shortly became known collectively as "The Boat Co." The partnership was not formalised until 1805, when they adopted the title of the "Brecknock Boat Co."[37] For a number of years they held an effective monopoly of the coal and lime trade on the canal.

In October 1794 it was ordered that "two active labourers" should by employed to keep the railroad in order, but by the following May it had been found that it was beyond their capabilities, and Knowles was instructed to advertise for a contractor to undertake the work. At the June meeting James Powell and John Arnold agreed to keep the line in order at £56 a year payable quarterly subject to a satisfactory report of their work, the sum of £16 to be withheld until the end of the year "by way of security for the due performance". In the event of their work being unsatisfactory, the B&A was to be at liberty to employ labour to put matters right at their expense. The B&A played fair with their new contractors, and the contract was not put into effect until 24 August 1795, when Knowles had put the whole in good condition.

The early days of the Nantyglo Ironworks were dogged by disagreements between the partners, R. S. Harford and Thomas Hill, and when Harford announced that the furnace was ready to be blown in, Hill withdrew his support, stating that too much of his capital had been spent.[38] It was about this time, June 1795, that Harford requested a branch from the Clydach Rail Road to the works. Rumours of the state of affairs must have been heard by the committee members, and the Clerk was instructed to inform Harford that they did not feel themselves justified in making it at the time as:

> " . . . it is currently reported that they do not intend to carry on the Nantyglo Works longer than while the present stock of materials is working out but as soon as there is any certainty that the Nantyglo Works will be regularly carried on or an engagement entered into for payment of Eight per cent on the sum laid out the Committee will pay due attention to Messrs. Harford's request".[39]

How right they were! By early 1796 the works was idle, and remained so for several years.

John Knowles' report of 28 April 1796 on the state of the B&A works gives a very useful summary of the progress that had been made by that date. The parts with which this present chapter is concerned are as follows:

> "The Main Line of Rail Road leading from Langroyney to Rhyd y Blew is Seven miles and Two Furlongs and is all finished and in complete Repair except the Bridge over the Usk which fell in February 1795 and has not been since rebuilt.
>
> There is erected on this Line of Road a Toll House with a Weighing Machine complete [SO.2418 1437] and another Toll House near Cwm Pucka [Cwm Pwcca approximately SO.211 124] and a Building near Penybont [SO.245 148] for a Carpenter's Shop, Black Smith's Shop and Warehouse covered in and nearly finished . . .
>
> The Branch of the Rail Road leading to Gelli Felen Colliery is Six furlongs and quite finished . . .
>
> The Branch of Road leading to Nantyglo is Five furlongs and is formed and stoned but not laid with Rails.
>
> The Branch of Rail Way leading to Lloyd-Coed Colliery is Two furlongs and is complete.
>
> The Branch of Road through Tyr Methig land is Five furlongs and was intended to form a Junction with the Trevil Road near Ebbw Furnaces; this branch is formed and Part of it stoned. The several Roads before mentioned . . . are all connected with the Main Line and that likewise forms a Junction with the Trevil Rail Road and thereby opens an easy communication with several Rail Roads made by the Proprietors of the Monmouthshire Canal and other Public Bodies."

He continued that the cost to date including £2,400 for the initial surveys, before the Act was obtained, and £1,200 for the Usk bridge, amounted to £29,000. From this figure must be deducted the £2,110 spent on the Llam-march Railroad, to arrive at a final figure of £26,890 for the Clydach Rail Road and its branches. Tonnages on goods carried in the year ended March 1796 had amounted to £539, but until the canal should be opened the full utility of the railroad could not be realised. He concluded with an inventory of the company's property which included at that time "23 Rail Road Waggons".[40]

Prepartations for cutting the canal were being made throughout 1796, and in October it seems to have been realised that no thought had been given to forming a connection between the railroad and the canal. Further land was accordingly purchased at Gilwern for a short branch from the main line to Glangrwyne into what was later known as Clydach Wharf. In May 1797 the necessary rails were ordered from the Clydach Ironworks at £7 12s. 6d. per ton, and they were paid for in August. The branch was probably completed in time for the opening of the canal to Llangynidr in early January 1798.[41] The Clydach Wharf was used by the Boat Company, who seem to have taken over right away, probably laying in stocks ready for the opening of the kilns the B&A were building for them at Llangynidr.

After the Nantyglo Works was closed in 1796, R. S. Harford continued to live at Nantyglo, and to work the Llwydcoed and Rhydw collieries (SO.179 110 and SO.119 124), probably supplying Walter Watkins at Glangrwyne coal wharf once the Boat Co. was occupied with trade on the canal. From correspondence and the canal minutes, it appears that his traffic was not particularly welcomed by the B&A, as it seems to have been largely carried in road waggons, which, in the case of Rhydw, appear to have been loaded by means of a chute spanning the railroad. This chute was considered a nuisance even before the collapse of the Nantyglo concern, due to the amount of rubbish that spilled onto the railroad;[42] and in 1797 the committee agreed to spend £7 10s. to build a wall to protect the railroad at this point, provided that Harford undertook to remove any of his rubbish that might fall into the river Clydach, and possibly threaten the railroad.[43]

Once the canal was partly open, it is likely that there was little traffic on the line from Gilwern to Glangrwyne, and in mid 1799 Harford offered to purchase this part from the B&A, stating that he was not interested in keeping the rails. As early as 1793 the B&A committee had suggested that a stone-road should be made alongside the railroad, but this had been negatived at the next meeting.[44] Harford's offer was turned down on 9 September 1799, and he was told that his road vehicles were no longer to have the use of the railroad. He protested that his coal would be " . . . entirely shut out from the market the other ford at Llangrwyney being so bad as to be impassable . . . "[45]

The committee's response was to order Knowles to put " . . . an effective stop to any other Carriage than Rail Way Waggons passing upon any part of the Rail Roads of this Company and particularly that part of the Rail Road between the River Usk and the Coal Wharf at Aberclydach . . . " and ordered the solicitor to enforce the agreement to keep the river clear at Rhydw.[46] On 17 October the assembly reinforced the order by passing a by-law that all vehicles should be fitted with wheels of a width of " . . . 4½ inches within the crest and fitted to move and keep up on the Rails . . . " Harford again protested on 31 October that not only would he be put competely out of the market and compelled to stop Llwydcoed Colliery, but that the "country" would be deprived of a source of its coal supply; he considered he had done all that was incumbent on him in respect of Rhydw, and pointed out that there were other parties with mining interests there.[47] By 1802, a new partnership had been formed to re-open the Nantyglo Works, and Harford had moved on to take over the Ebbw Vale Ironworks.[48]

III

By March 1799, the canal was completed from Gilwern to Talybont on Usk, and progressing towards Brecon. Probably the only carrying at the time was for the Boat Co., who were thanked publicly at the assembly of 25 April, "for their exertions in establishing a trade".

At their March meeting the Monmouthshire Division committee dropped a bombshell when they proposed that contracts for canal construction towards Pontymoil should be abandoned and a railroad substituted, and suggested that Benjamin Outram should be called in to examine and report on both the canal and the railroad. At first the Breconshire Division committee demurred, but later agreed with the suggestion, and Outram was engaged on 20 June, at a joint committee meeting.

He must have had a busy ten days, as his report dated 1 July 1799 covered some eighteen miles of canal, eight miles of railroad and a survey of an alternative route for the latter.

Always a staunch proponent of plateways, the opening words of the report in respect of the railroad come as no surprise when he writes:

> "The Railways have been laid out agreeable to the old system and are expensive to work and troublesome to keep in repair. They might be worked to greater advantage by issuing lighter Waggons with smaller wheels turning on the axles, but if instead, others were substituted on the system I have introduced, the Expence of Conveyance might be reduced Three-Fourths, and the Expence of Repairs in proportion."

and continues to advocate that all lines should be altered to plateways on rails weighing from thirty-two pounds to forty-four pounds each, carrying waggons of from one and a half to two and a half tons capacity.

He suggested a complete re-routing of the line of the railroad by taking it from the junction of the line to Llwydcoed somewhat nearer to what is now the main road through Brynmawr, before passing down the south side of the Clydach valley as far as the Llam-march valley at an inclination of 1:108, followed by an incline of 400 yards at 1:8. From the foot of the incline the line would be continued at a descent of 1:54, to terminate to the south of Clydach aqueduct, Gilwern, some twenty feet above the level of the canal, where he suggested the establishment of a stock yard for coal and limestone, with chutes to load the boats. He considered that as far as the incline was concerned, no engine would be required if the "downgate" traffic was at least three times as heavy as the "upgate". He specified the use of a chain on the incline in preference to a rope.[49]

There is no record in the canal minutes of any discussion taking place on the report, either by the proprietors or in committee, but there is a gap of about two years prior to July 1802, for which the committee minutes are not available, and any discussion would probably have taken place before that date.

A special assembly on 18 February 1802 decided to go to Parliament for powers to raise more money, and to " . . . make a new Dram Road up the south side of the Clydach valley . . . and . . . to discontinue the Present Road . . . " An Act was obtained on 3 May 1804, but no attempt was made to alter the railroad. This has been confirmed by fieldwork, and had an alteration been made, through running with the Trevil Rail Road would have been impossible. That no alteration was made is confirmed by the B&A published accounts for March 1843,[50] which show the following items of expenditure:

Repairs to Rail Road	£47 3s. 0d.
Repairs to Tram Road in Brecon ...	£1 10s. 0d.
Tram Plates and iron Rails	£28 0s. 0d.

In April 1800, Edward Kendall approached the canal company with a request for a railroad to be constructed to a colliery he proposed to open at Llwynypwll Farm (SO.2033 1185), but with the general shortage of funds that the company was then experiencing, the request was turned down, permission being given to Kendall to make the line for himself. The engineer, at the time Thomas Cartwright, was instructed to make any requisite deviation in the railroad to effect a junction, but at Kendall's expense. Later in the year, the colliery was sub-let to Wilkins, Lloyd, Powell and Peirce, of the Brecknock Boat Co.[51] The point at which the later Merthyr, Tredegar & Abergavenny Railway crossed this railroad (SO.1913 1191), was known as "Brecon Boat Siding" until the railway was closed in 1958.

A fresh partnership was formed in 1802 to re-open the Nantyglo Works, again with Thomas Hill of Blaenavon as one of the principals. Doubtless inspired by the iron company, a William Williams, of Trosnant, Nantyglo (formerly the dwelling of R. S. Harford), applied, on 25 April 1805, for a railroad from the Clydach Rail Road, "at or near the Upper Toll-Gate on the

Clydach Valley. Site of the Clydach Rail Road in 1960, near Black Rock Quarry before construction of the Heads of the Valleys Road.

Mountain", to a point near the furnaces. Scenting possible trouble, the B&A refused to make the line but gave Williams permission to construct the line for himself. It did not materialise, however, and by 1808 the works was again idle, due to disagreements between Hill and the other partners. Once again, the canal company had been wise, or had, perhaps, heard of the difficulties experienced by the Monmouthshire Canal in obtaining its tonnages from the Nantyglo Co.[52]

The Nantyglo Works was taken over in 1811 by Joseph Bailey, a nephew of Richard Crawshay, of the Cyfarthfa Works, Merthyr Tydfil, and Matthew Wayne. The B&A displayed a marked change in their previous attitude to requests for communication with the works when the new partners asked for permission to lay rails on the stone road leading to it, acceding immediately subject to an annual payment of £1 1s. and the right to take tonnages should any traffic, other than that of the iron company use it.[53]

Six months later, on 13 November, with the prospect of the B&A reaching Pontymoil in a matter of a few months, Bailey & Wayne put a fresh proposition to the canal proprietors ". . . which will certainly be much to the profit of your concern and to us a convenience of not having so many different kinds of carriage in getting to market . . . "[54] They suggested that the B&A should make the mile of line from the main line to Nantyglo, and permit them to use both that and the main line to Gilwern, free of charge, in return for which all the output of the works would be sent out via the canal. In a postscript to the letter containing the proposals, Bailey & Wayne added ". . . We expect the Nantyglo Works will make up to 100 tons of iron per week, and in getting to Newport will go $13\frac{1}{2}$ miles upon the Brecon and Abergavenny Canal. It will produce to the Canal at 3d. per ton per mile £877 10s. a year".

The B&A was in general agreement with the proposition, but stipulated that the railroad from the main line should be made by the iron company, and provided that this was done at or under a cost of £600 they should be at liberty to deduct the cost from their tonnages on the canal, and once reimbursed in this manner, the branch of railroad should be vested in the canal company. Bailey & Wayne were to be liable for all maintenance of the branch, and were to pay the B&A as though all their output had been sent along the canal even if some part of it might be exported by other means. These terms were agreed and the contract was finally sealed on 13 December 1811.

The B&A was opened to Pontymoil on 7 February 1812, and in preparation for their traffic Bailey & Wayne applied for part of Llanelly Wharf to be set aside for their sole use in stacking iron, and the B&A engineer was instructed to ascertain if land could be procured for this purpose. By December 1812, the Llwydcoed branch was out of use, and possibly had been since the departure of Harford in 1801, and Bailey was given permission to use the rails lying idle there on condition that he undertook to replace them with others of a similar quality if called on to do so. It would seem that a great deal of preparatory work was needed at Nantyglo before production could start, and it was not until 17 January 1814 that the B&A was notified of the completion of the railroad, and the Agent instructed to allow the iron company £600 credit on their tonnage account.

The Brecknock Boat Co. were granted similar terms to construct a short branch to their new colliery near Clydach Bridge (SO.1986 1213) on 19 November 1812.[55] In their case the cost was limited to £75, to be deductible from their tonnages, and when paid for the branch was to be vested in the B&A; assuming costs to have been on the same basis as those of the iron company, this line would have been about 220 yards long. No date for its completion has been found, but in March 1814, it was discovered that the BBCo. had been undercredited £9 in respect of the cost of blocks and sleepers, and they were allowed further credit.[56]

By April 1813, trade on the canal and railroad had increased so much, that it became imperative to provide additional wharf space at both Llanelly and Brecon, and the Agent was instructed to find more land at both places. In June he was authorised to enter into negotiations with landowners, and in October a sub-committee was appointed to examine the site at Llanelly and to consult the BBCo. as to its suitability. Authority to purchase the land at £150 an acre was given in December.

The difficulty of ascertaining the exact date of implementation of decisions taken by the B&A is well illustrated by what happened in respect of this wharf. In April 1814, the solicitor was authorised to purchase the land for the wharf and a connecting branch railroad. In October, a lease was concluded with Bailey & Wayne for a wharf of two boat-lengths, and the space to construct a road from it to the public road, all of which the iron company undertook to fence at their own expense; later in the same month a request was received from the coal and lime traders for a railroad to the new wharf. The Agent was instructed to complete it without loss of time. Nothing more is heard of the new wharf or of the railroad branch until 13 August 1817, when it was reported to the committee that " . . . the traders on this canal are greatly in want of wharf room at Llanelly and cannot carry their produce along the line of the canal . . . without the same being immediately provided . . . " The Agent was instructed to complete the wharf " . . . as instructed on 12th March 1812."! He was also told to contact the BBCo. to inform them that it was intended to carry the railroad branch at the back of their premises at Clydach Wharf. The date of opening of the branch has gone unrecorded.

The depression following the end of the Napoleonic Wars was felt very severely by the South Wales iron trade. In an effort to reduce expenses Joseph Bailey suggested to the assembly of 19 October 1815 that his imported iron ore should be carried free on the canal and railroad in return for his putting his whole output on the canal. His request was granted, and the committee was empowered to grant similar facilities if asked for by other iron manufacturers, but reserved the right to withdraw the concession should it work out to the disadvantage of the B&A. A year later it was resolved that provided it was afterwards shipped on the canal as far as Pontymoil or the equivalent distance towards Brecon, all iron should be carried free on the Clydach Rail Road.[57]

Further schemes for reducing the tonnages on iron were the subject of letters received from Kendall & Bevan, of Beaufort Ironworks, and Harford, Crocker & Co. of Ebbw Vale on 4 March 1817, both proposing that if the B&A would reduce the tonnage on iron to 2d. per ton per mile, and permit branches from the Clydach Rail Road into their respective works, on which tonnages would be charged, they would both send all their produce to Newport on the B&A. Approval in principle was given at the assembly of 24 April, and the tonnage reduced as requested, but it was laid down that Harfords should not spend more than £1,000 on their branch, and limited Kendall & Bevan to £500 for theirs. Any variation in the arrangements was to be subject to six months notice on either side. The following are the calculations on which the arrangements were based.[58]*

"Estimate of the advantages proposed by the lowering of Tonnages upon Iron from 3d. to 2d. per ton per mile.

Clydach Iron Works 14½ miles 4,000 tons at 2d.	£483 6s. 8d.
Nantyglo Iron Works 18 miles 6,000 tons at 2d.	£900 0s. 0d.
Ebbw Vale Iron Works 19½ miles 3,600 tons at 2d.	£585 0s. 0d.
Beaufort Iron Works 19½ miles 3,600 tons at 2d.	£585 0s. 0d.
Total	£2,553 6s. 8d.
Tonnage of iron upon the canal for the last two years at 3d. per ton per mile = £3,499 17s. 6d.	
One year produced	£1,749 18s. 9d.
Balance in favour of the Canal Company lowering the tonnage to 2d. per ton per mile	£803 7s. 11d.

This reduction in tonnages, naturally, did not please the MCC, who were thereby forced to reduce theirs in accordance with clause 108 of the B&A Act. On 26 June 1817, Harfords informed the B&A that the MCC proposed to charge them 5d. per ton per mile on that part of the journey over MCC water, and whilst the B&A considered it was not their place to interfere with the actions of the MCC concerning their own canal, they thought it was an active intervention in the affairs of the B&A. On 16 January 1818, the MCC decided to take legal action against Harford, Crocker & Co. and Kendall & Bevan, " . . . ironmasters, who have deserted the Monmouthshire line of Road and Canal for the Brecon line, to compel the payment of Tonnages due to this Company according to the rates fixed by the Monmouthshire Canal Act . . . "[59] Skillful manoeuvres by Harford's solicitors had the case adjourned until June 1821, and in the meantime the ironmasters refused to pay any tonnages to the MCC, until, in desperation, that company asked for payment of 2d. per ton without prejudice to any pending action. Eventually, when judgement was given, it went against the MCC who were forced to repay such tonnages as had been overcharged.

* From the mileages quoted it would appear that the arrangements for free carriage on the railroad were to be cancelled.

Clydach Rail Road approaching Brynmawr, 1960.

John Lloyd[60] suggests that the Beaufort Works commenced sending their traffic to the B&A by carrying their iron to the top of Beaufort Hill on horse-back, but it is more probable that it was carried on the Trevil Rail Road branch from Beaufort to Cwmgwybedog, as far as the point at which it crossed the Clydach Rail Road, about half way up Beaufort Hill, and was there transferred to the B&A line either by transhipment or by transferring the waggons to the other line which was possible as both lines were on the same gauge.

No trace has been found of any connection with Ebbw Vale having been made by Harfords and in view of the fact that they gave six months notice to terminate the agreement on 19 March 1819[61] it is unlikely that it was made. An alternative route would have been available by the TRR branch to the works which connected with the Clydach Rail Road at Rhyd y Blew.

Matthew Wayne left the Nantyglo Works about 1819, and established his own furnace at Gadlys, Aberdare. His place at Nantyglo was taken by Joseph Bailey's brother, Crawshay Bailey.[62] In August 1819, the brothers were given permission to erect a warehouse on the new wharf at Llanelly at a maximum cost of £120, which the B&A undertook to purchase from them when completed, the Baileys then to pay a rental of eight per cent of the cost. In the event the B&A bought it from them in October 1821 for £80, on the understanding that possession would be given in twelve months time, when it was to be adapted to house the Llanelly wharfinger. Probably by this time the Baileys were thinking in terms of their own tramroad to Govilon, which was built about two years later. Another ironmaster, George Brewer of Coalbrook Vale

Ironworks also requested warehouse accommodation at Llanelly, on 17 January 1820, and presumably forwarded his iron on the B&A.

Kendall & Bevan, who had not yet made the connecting railroad to Beaufort, approached the assembly on 27 April 1820 with fresh proposals to connect with the Clydach Rail Road near Wain Dew, under which they were to construct the line at the expense of the B&A and to pay eight per cent of the cost as annual rent during such times as they had three furnaces in operation, but should they be using fewer furnaces the rent should be reduced to five per cent—a wise precaution in a time of declining trade. Should they cease to send out their iron by this route, they undertook to repay the full cost of construction less any amounts previously paid as rent. The B&A agreed, subject to a limit of £500 being spent on the project, and to the approval of the line by the Agent when made. This line was made and left the Beaufort Works (SO.169 111) by an incline of about 700 yards to join the Clydach Rail Road at SO.1739 1149, on top of the later Beaufort tunnel of the MTA. Shortly after 1833, the incline was incorporated in the tramroad connecting the Nantyglo and Beaufort Works under the Bailey brothers.

Kendall & Bevan were not the only ones thinking of economies. A sub-committee of the B&A, appointed to examine possible ways of saving money reported on 19 October 1820. They were critical of the company's book-keeping and suggested that the method employed by the Oxford Canal should be introduced, and at the other end of the scale that the mole catcher should be instructed to catch rats also, at of course, the same wage. With respect to the Clydach Rail Road they reported that the weighing machine was defective, and was too short to weigh bar-iron; the toll houses were in bad repair, and they considered the £500 or so that it cost to keep the railroad serviceable to be objectionably high. They recommended that the by-laws should be more stringently applied, one half of any penalty charged to an offender should be paid to the informant. On matters outside the direct control of the company, they recommended that the Baileys should be asked to strengthen the dam of their pond, as its weakness was endangering the Mountain Toll House (approximately SO.182 113). Two rather surprising points emerge from this report. When purchasing the land for the new wharf at Llanelly, no right of way into the site had been acquired: and in spite of very high arrears of tonnages on both railroad and canal, no effort had been made to take security from freighters having credit accounts.

Further recommendations were that the committee should inspect the whole of the canal and railroads at least once every six months, and that the maintenance of both branches of their business should be in the hands of contractors.

In April 1821, the Agent was ordered to exercise the strictest control over the expenditure on canal and railroads, and to cut the wages of all unskilled staff. This was followed in October 1822 by reductions of between ten per cent and fifteen per cent in all salaries, any who protested were to be liable to immediate dismissal.

The B&A monopoly of the Nantyglo traffic was broken by the construction by the Baileys of their own tramroad to Govilon from the works, which was completed in 1821. Although the canal was still used to reach Newport, two miles of canal tonnage was lost to the B&A, in addition to the loss of traffic on the railroad. In July 1828, the committee received news that the MCC had completed their tramroad from Aberbeeg to Coalbrook Vale. That the Baileys had made a connection thence to Nantyglo completed the gloomy picture, as there was now only one works, Clydach, which was completely reliant on the B&A. A meeting was called of the ironmasters who would now have direct transport via the MCC, as a result of which the committee recommended a reduction in tonnage on iron from 2d. to 1½d. per ton per mile, which the assembly of 16 October confirmed, after the proprietors had received a circular letter informing them that unless the reduction was made, the iron freighters would be unable to continue to use the canal. Two years later, a further reduction to 1d. per ton was made in an effort to retain the traffic.

It is doubtful if the line to Glangrwyne ever brought much traffic to the canal, and, indeed in 1811 the B&A had refused to construct twenty yards of track to carry the railroad actually to the edge of the ford.[63] There was thus no panic when, in October 1824, it was realised that the lease of that works was about to expire, and the B&A appointed two of their committee to superintend the staking out of the company's property to the north of the Usk, including the road built by them from the river to the village of Glangrwyne. Walter Watkins, who had run the company's "land sale" depot at Glangrwyne, was dead[64] and it is probable that by 1824 the line from Gilwern to the ford was practically disused. By 1833, the B&A agreed to the parish of Llanelly using the line for a road at a nominal rental of 1s. per annum.[65] It is likely that more road traffic than rail was then passing, the gatekeeper at the Clydach Gate on the Crickhowell-Abergavenny turnpike, which stood near the present Corn Exchange Inn, complaining that the turnpike was so badly out of repair, that even foot passengers preferred to walk on the line of the railroad, and cross the ford.[66] In October 1838, it was ordered that the rails be taken up and sold, and "... such useless part of the Road ..." was let to John Powell, who then held the Clydach Ironworks, at £2 2s. a year. In 1839, it was referred to by the Breconshire Turnpike Trust as a highway.[67]

By 1830, little or no iron was being carried on the railroad or on the canal, and the main source of income to the B&A was the coal, coke, and limestone traffic of the BBCo. By 1833, the Beaufort Works was in the hands of the Baileys, and iron made there was sent to Nantyglo for processing, and exported on the MCC tramroad through Aberbeeg. Ebbw Vale Works had the use of the MCC's Beaufort Tramroad and the Sirhowy Tramroad, and the Clydach Works had their own line to the canal which they leased from the B&A. A fair indication of the state of affairs on the railroad can be ascertained from the B&A published accounts for September 1836, which show that in the previous half-year, 9,922 tons of coal and coke had been carried yielding tonnages amounting to £321 1s. 11d. and 2,145 tons of lime and limestone yielding £22 4s. 9d. Only half a ton or iron, yielding 2d. in tonnages had passed in the same period. Generally this was to be the future picture of traffic on the railroad.

The BBCo. had been authorised in 1812 to construct a branch railroad to their Clydach Colliery (page 72), and under the terms of their agreement with the B&A this had been vested in the canal company when completed. A sub-committee appointed in July 1836 to examine the various canal properties used by the BBCo. was empowered to contract with them for repairs on the branch. It would appear that they were unwilling to undertake this, as in October the Agent was instructed to put the whole railroad in order, and to substitute stone blocks for iron sleepers, with no mention of the branch being excepted.

By 1839 it was considered that the line to the west of Brynmawr was "utterly useless", and in July 1841, the Agent undertook the maintenance of the canal and railroad for £2,000 a year, but it is doubtful if much of this was spent on the railroad; in 1843 repairs to the line were only £47 3s. in the half-year ending in March.

When the question of repairs next came before the committee, in April 1850, it was decided to advertise for tenders in *The Times,* the *Railway Times* and the *Bristol Mercury,* but by July only one offer had been received, and that was considered to be extortionate. In October 1851, John Lloyd of the BBCo. offered to contract for repairs at £300 less than the average cost, or £1,602 per annum. The offer was rejected and the committee resolved to re-advertise. In November the tender of Richard Pruett to effect repairs to the canal and railroad for £1,445 a year was accepted.

The terms of his contract show that it was still considered that the railroad extended from " . . . the ford on the river Usk opposite Langroiney Forge to two spots near to and beyond Beaufort Iron Works . . . ", despite the statement in 1839 that beyond Brynmawr it was utterly useless, and the order of October 1838 that the rails should be lifted between Gilwern and the river. For the canal part of his contract, Pruett was bound to hire a boat from the B&A, and for the railroad he had to hire "the Company's two Rail Road Waggons" and to keep them in repair; also he had to purchase all their stock of spare rails, sleepers, blocks, etc. It is interesting to note that the contract bound Pruett to replace any broken rails with new ones of wrought iron. Rails were to be laid on stone blocks where practicable, but where the bed rock was too near the surface for them to be bedded in, wooden sleepers of "sound heart of oak" were to be used.[68]

Pruett's contract was renewed in December 1856 for a further five years, but in the reduced sum of £1,300, and its scope was extended to include the Watton Plateway in Brecon, and the repair of the company's property including houses, toll houses and warehouses. Before a further renewal was granted in 1861, the company's works were examined by a Mr. Kirk, who reported that in his opinion it would be necessary to spend £233 10s. to bring the canal to a suitable standard of maintenance, and that £114 should be spent on the railroad.[69] The canal repairs were sanctioned, but those of the railroad were "left for future settlement".

By September 1860, the MCC (now the Monmouthshire Railway & Canal Co.) had adapted all its tramroads to fully fledged railways, with locomotive haulage, providing easy through transport to Newport for all the ironworks except Clydach, and the attractiveness of the B&A route had entirely faded as far as the ironmasters were concerned. The BBCo. still used the railroad and the canal, but the accounts for 1851–59 show that tonnages received on the railroad averaged only £612 a year over the period. How little other traffic was being offered can be judged from the fact that on 17 October 1861 the contractor was given permission to occupy the Llanelly machine-house rent free, its original purpose having been entirely lost.

The BBCo. complained of the state of the line in October 1861 and the contractor was ordered to put it in the state it had been when he originally took it over. Nevertheless, the BBCo. was told to reduce the loading of its waggons, and was allowed £20 a year for "roughing it". There was a surprising decision taken in July 1863, when an outlay of £100 on rails was authorised, and the contractor was ordered to lay them as soon as possible. It was, however, to be his swan song as far as the railroad was concerned, as the maintenance of the line was taken out of his hands in August, and his contract was reduced by £72 a year. Possibly the repairs were then undertaken by the BBCo. until that company changed hands in September 1866,[70] after which it has proved impossible to obtain further information.

To the west of Brynmawr, where the railroad had been unused for some years, various parts were let to persons who had "squatted" on the line at 1d. per square yard for fifty years in 1863.[71]

With the opening of the MTA and the Brecon & Merthyr railways, taking the coal traffic, the affairs of the B&A continued to worsen, and in 1864, after many endeavours to sell the concern to any railway company willing to purchase it, the offer from the MR&CC to buy the whole concern was accepted. The necessary Act was obtained and the transfer was effected as from 29 September 1865. The committee continued to meet to wind up the affairs of the company, and on 26 April 1866, the whole railroad to the west of Clydach Bridge, Brynmawr was sold to J. & C. Bailey for £300, the low price giving an idea of how little its loss was likely to affect canal prosperity. The final meeting of the B&A committe was held on 13 June 1867 to pass the last accounts.

Route

From the village of Glangrwyne (SO.240 163) to the site of Glangrwyne Forge (SO.2395 1613), there is no record of rails having been laid, and it is probable that this stretch was always the stone road ordered to be constructed by the committee on 15 April 1794, but there is no doubt that rails were originally laid from the forge and across the bridge over the Usk. There was probably a weighing machine between the forge and the bridge.

The present bridge is on the site of the original wooden structure (SO.2405 1606). From its eastern end the railroad turned south, on the line of the existing minor road, the first twenty chains of the road being built on the embankment made under Jonathan Gee's contract. Aberclydach Bridge, by which the railroad crossed to the west of the Clydach river, at SO.2456 1508, is also on the original site, and is probably the original constructed by Prichard.

Continuing on the line of the minor road, and passing the Corn Exchange public house (SO.2452 1486) the railroad crossed the Abergavenny to Crickhowell turnpike road (A4077) and entered Church Road, Gilwern, but almost immediately forked southward away from it. At this point, the 1902 edition 25″ O.S. shows that there was a wharf, and that the railroad was *in situ* to this. Running south to SO.2450 1462, the railroad then turned south-west to pass under the canal at SO.2448 1450, and ran to Maesgwartha Road at SO.2420 1439.

Llanelly Wharf extended from the kilns at SO.2422 1477 to the road to Llanelly village at SO.2433 1463. The railroad leading to the kilns lay on a rising embankment taking it to the level of the charging mouths; the tapping hole was on the level of the canal bank. This embankment can still be seen passing in front of Auckland House (SO.2425 1465). Leaving the wharf at SO.2431 1462, the railroad crossed Church Road and passed immediately into Maesgwartha Road which it followed to join the line from Glangrwyne, passing at SO.2433 1463 the road entrance to Clydach Wharf.

Clydach Wharf, which appears to have had a fan of sidings to accommodate the traffic to the Brecknock Boat Co., stretched from the road to Llanelly at SO.2433 1464 to the head of the dock at SO.2434 1449. This dock was originally the point at which the canal feeder from the river Clydach entered the canal, and the railroad entrance to the wharf is marked by massive gate-posts. It joined the line from Glangrwyne at SO.2430 1446.

A weigh-house was situated at SO.2418 1437, the place decided on by Jeffreys Wilkins and Walter Watkins in 1794.[72] On 29 June 1795 orders were given for a dewlling-house to be erected on the site and on 18 October 1810, it was represented to the committee as being too small, as the books and the machine were all inside the dwelling, and it was ordered that "a kitchen or common room be built adjoining" at a cost of under £40. The weigh-house still stands, rejoicing in the name of "Machine House", but it is difficult to determine which parts, if any, are of the original construction.

The railroad continued along the road to pass through the hamlet of Maesgwartha, passing the site of Hanbury's furnace at SO.233 138, to the crossroads at SO.2297 1365, to which point there was probably a plateway connection from the Clydach Ironworks. At SO.2269 1318, the railroad was joined, about 1809,, by the Abergavenny to Merthyr turnpike and the two roads ran alongside each other for some nine chains to SO.2260 1301. The conjunction of the two roads at this point caused the turnpike Trustees a certain amount of worry as they thought that:

> ". . . the Tolls arising from that part of the said Roads . . . may by much evaded by carrying-horses passing or pretending to pass up and down the said Rail Road . . . "[73]

The B&A acquiesced in the Trustees' request not to oppose protective clauses which they intended to include in their Act.

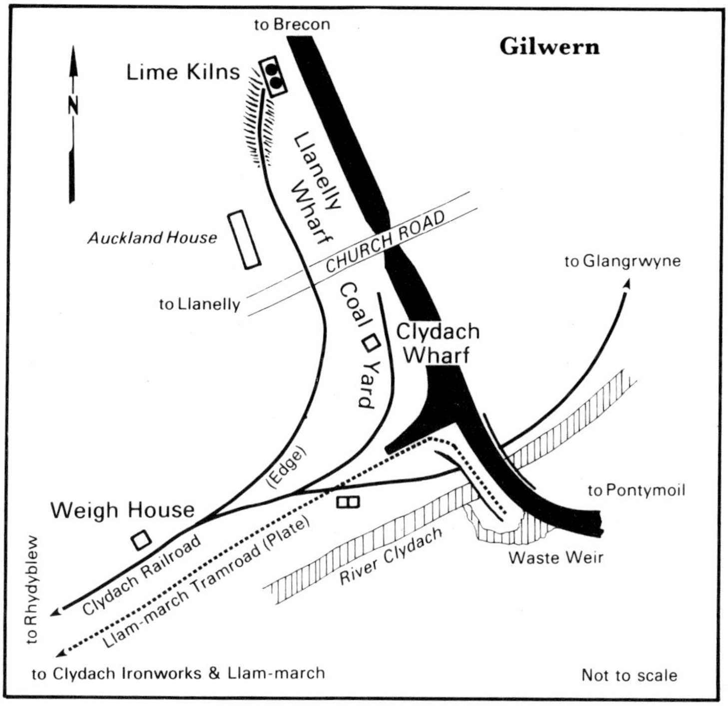

From SO.2260 1301, the railroad can be traced as an asphalted path running between the old main road and the Heads of the Valleys Road. This comes to an abrupt end at SO.2210 1275, where the construction of the latter road has obliterated all traces. How, in former years, the railroad was regarded as a right of way can be judged from the 1970 printing of the 6″ O.S., which shows two schools and a chapel whose only access is by means of the path.

The railroad track can again be picked up for a short way (200 yards) from SO.2160 1260 to SO.2151 1250, passing to the north of the *Old Drum & Monkey Inn* (SO.2154 1256). An interesting feature that has been lost due to the road works, was the railroad entrance to Daren Ddu Quarry at SO.2176 1268, passing under the erstwhile turnpike road, by a tunnel of about twenty yards in length.

The next point at which the railroad track can be found is close to the site of the former Pont Harry Isaac (SO.2045 1232), which was another casualty of the Heads of the Valleys Road. It was here that the branch from Edward Kendall's Gelli Felen Colliery joined the main line. The colliery was at approximately SO.214 119, and the branch ran on the line of the road from SO.2139 1210, to cross the later MTA at SO.2183 1229, and then fell sharply by an incline to the bridge.

At SO.2022 1226, there still stands a culvert over the Nant Melyn which falls to join the river Clydach. The railroad at this point is a public footpath, and the local council has strengthened this culvert in a manner which has not improved its appearance.

Yet another casualty of the Heads of the Valleys Road was Clydach Bridge, SO.1965 1213, the site of which is under the large roundabout at the west of Brynmawr. The footpath on the line of the railroad from Pont Harry Isaac is, however, continued from the roundabout as a path leading to Alma Street, the two portions being connected by a subway under the roundabout. The path to the west of the subway passes the site of Hafod Pit, one of those worked by the BBCo.[75]

Entering Alma Street, Brynmawr, at SO.1949 1202, the railroad was joined almost immediately by the BBCo's. line from Llwynypwll at SO.1943 1197. A rail- or tram-road has been traced from SO.1936 1212, which was on the land of Llwynypwll Farm, but the lease from the Kendalls covered a wide area and it is probable that coal was worked at many points in the course of the years, making the actual situation of the Llwynypwll Colliery difficult to locate with any certainty.

Passing along Alma Street, the railroad took the lower of the two levels, and entered what appears to have been known in earlier days as "the deep cutting on Waun yr Lychen" (Waun yr Helygen—Moor of Willows, a name given to the part of Brynmawr which now comprises the southern half of the town). To the east of this cutting, the railroad was joined by a line constructed by Benjamin Price from his colliery at Ffosmaen,[76] to the south of Nantyglo. This cutting was bridged at its western end by the Daren Disgwylfa line of J. & C. Bailey, authority for the construction of the bridge being given on 30 August 1819, by the B&A committee. The eastern end of the cutting was bridged at approximately SO.1909 1168, by Bailey's railroad to Llangattock, about 1830. Old inhabitants of Brynmawr will state that the Clydach Rail Road was in a tunnel under the Square, but this must have been a much later development as the whole of Waun yr Heligen became built over. As a result of this development the railroad is, at the present time, lost until SO.1853 1143 is reached, adjacent to the now dismantled line of the Brynmawr & Western Valleys Railway.

From here it passed to the south of the pond at SO.183 114, which was one of the reservoirs supplying the Nantyglo Works, but is now considerably smaller than when it was operational. What was usually referred to as the "Toll Gate on the Mountain" probably stood to the south of this pond between its most southerly point and the branch to Llwydcoed at SO.1819 1132. A plan of the railroad drawn in 1795[77] shows that the original stone road to Nantyglo Works also started at this point, and it is possible that the road shown on modern maps leading from here to Winch's Row (SO.1859 1100) represents the line of the stone road, and the railroad to the works completed by Bailey & Wayne in 1814.

The Llwydcoed branch has been obliterated by later workings, but probably terminated at SO.181 111.

The route of the main line can now be followed as a footpath between wire fences to SO.1760 1155, where modern housing takes over, and for the next three-quarters of a mile one has to rely on inspired guesswork. At SO.1742 1148, above the now blocked tunnel of the MTA, is probably the point at which the incline to Beaufort Works left the main line, which seems to have run between the houses in Frost Road, Beaufort, and the MTA as far as Bryngwyn Road, SO.1697 1166, where it turned to the south-west to run behind the theatre, and the church at SO.1661 1156, crossing Raglan Terrace (formerly the line of the Trevil Rail Road branch from Beaufort Works to Cwmgwybedog) at SO.1671 1158.

Following the line of the main road (A4047), the line of the Clydach Rail Road can next be identified at SO.1629 1160, where it entered the road to the north of the graveyard attached to Carmel Chapel, which it followed to its junction with the Rassa Rail Road at SO.1609 1154.

REFERENCES:

1. Rees, W. *Industry before the Industrial Revolution.* University of Wales Press, 1968.
2. Lloyd, J. *Early History of the Old South Wales Iron Works, 1760–1840.* 1906.
3. NLOW. Maybery 488.
4. ibid. 489.
5. ibid. 492.
6. ibid. 490.
7. ibid. 566.
8. ibid. 487.
9. ibid. 2463.
10. ibid. 486.
11. PRO. MCC committee minutes, 4 September 1792.
12. ibid. 2 July 1793.
13. NLOW. Maybery 491.
14. PRO. B&A committee minutes, 14 June 1793.
15. NLOW. Maybery 2474.
16. ibid. 493.
17. ibid. 2478.
18. ibid. 2479.
19. ibid. 496.
20. ibid. 499.
21. ibid. 4032.
22. ibid. 510.
23. ibid. 523–4.
24. ibid. 518.
25. PRO. RAIL/812/9.
26. PRO. B&A committee minutes, 15 October 1794.
27. ibid. 26 June 1794.
28. NLOW. Maybery 572.
29. ibid. 6385.
30. ibid. 529.
31. PRO. B&A committee minutes, 20 May 1795.
32. NLOW. Maybery 566, 570 and 6386.
33. ibid. 296.
34. Lloyd, J. op. cit.
35. NLOW. Maybery 530.
36. The long ton consisted of 21 cwt. each of 120 lbs. (2,520 lbs.), as against the Statute ton of 20 cwt. each of 112 lbs. (2,240 lbs.).
37. GRO. D.591/11.61.
38. Lloyd, J. op. cit.
39. PRO. B&A committee minutes, 4 June 1795.
40. Brecknock Museum, Brecon, Maybery collection Vol. 1. fo. 9.
41. PRO. B&A committee minutes, 4 December 1797.
42. ibid. 29 June 1795.
43. ibid. 17 October 1798.
44. ibid. 27 August 1793.
45. NLOW. Maybery 3816.
46. PRO. B&A committee minutes, 16 October 1799.
47. NLOW. Maybery 3924.
48. Lloyd, J. op. cit.
49. NLOW. Maybery 383.
50. NLOW. Tredegar 80/808.
51. GRO. D.591/11.61.
52. PRO. MCC committee minutes, 28 February 1805.
53. PRO. B&A committee minutes, 16 May 1811.
54. ibid. 14 November 1811.
55. Location of colliery is shown on a plan of the land utilised for Bailey's Tramroad. GRO. Man/2/2–278.
56. PRO. B&A committee minutes, 28 March 1814.
57. PRO. B&A proprietors minutes, 17 October 1816.
58. PRO. B&A committee minutes, 4 March 1817.
59. PRO. MCC committee minutes, 16 January 1818.
60. Lloyd, J. op. cit.
61. PRO. B&A committee minutes, 21 October 1819.
62. Lloyd, J. op. cit.
63. PRO. B&A committee minutes, 16 May 1811.
64. ibid. 16 October 1822.
65. ibid. 25 April 1833.
66. NLOW. Maybery 1505.
67. ibid. 1822–3.
68. ibid. 1031.
69. PRO. B&A committee minutes, 25 April 1861.
70. NLOW. Badminton 7729.
71. PRO. B&A committee minutes, 15 October 1863.
72. ibid. 24 April 1794.
73. PRO. B&A proprietors minutes, 20 October 1808.
74. GRO. Man/2/2–278.
75. 1" O.S. Sheet 232 of Geological Survey of Great Britain (England & Wales), 2nd impression 1963.
76. PRO. B&A proprietors minutes, 29 April 1819.
77. "Plan of the Brecknock & Abergavenny Railroad from Langrwyney Bridge to Rhyd-y-Blew . . . " In possession of British Waterways Board, Gloucester.

8

Llam-March Tramroad

When establishing the Beaufort Ironworks, Edward and Jonathan Kendall, in common with many other ironmasters in South Wales, found their land holding was larger than could be worked economically by the one works. In partnership with Edward Frere and Thomas Cooke, Edward Kendall commenced to build an ironworks at Clydach in 1793, and in 1797 E. & J. Kendall sub-let the Clydach site to Frere and Cooke. The Duke of Beaufort, from whom the Kendalls had obtained the original lease, objected to the sub-letting, and in 1801 insisted on a fresh lease being drawn for all the land held except that actually being used for the Beaufort Works.[1] This necessitated a fresh sub-lease to Frere and Cooke, and Edward Kendall took the opportunity to leave the partnership. He did not, however, withdraw his capital but ensured a sound investment by imposing the high rental of £780 a year on the smaller concern for land on which he was paying annually £17 10s.[2]

The Clydach works was situated in the bottom of the valley, which for its whole length is very narrow. Whilst there was adequate water-power, the site was very cramped, leaving little room for the disposal of the scoria; iron ore and coal for coking had to be obtained from the hills around Brynmawr, some 600 feet above the level of the works. Probably the main inducement to Kendall to build the works in this location was the prospect of cheap transport offered by the B&A, but until this route to Newport became available in 1812, it was very badly placed for transport to its markets.

To finance the undertaking, Frere borrowed large sums from his two brothers against his and Cooke's guarantees, and further capital was injected in 1804, John and Walter Powell, the Brecon solicitors, each taking a one sixth share. Cooke retired in 1814, leaving Frere with a fifty per cent holding, with Walter Powell controlling the other fifty per cent comprising his own share and that of his deceased brother whose executor he was.[3] In 1820, Frere sold one-third of his holding to John Powell, son of the original John, and the Powell family acquired a majority holding in the concern. Frere retired in 1832, and the Powells took up his share, and remained in control of the works until they failed in 1861. The final blow came when the B&A obtained judgement against them in respect of £1,650 owed for tonnages, and the works was sold by the trustees in bankruptcy.

The company never really flourished, and was bedevilled throughout its existence by the large sums owed to the Frere family. For its day to day working capital it had to rely on borrowing from its bankers, Wilkins & Co. of Brecon, with the consequent addition to its interest charges.[4]

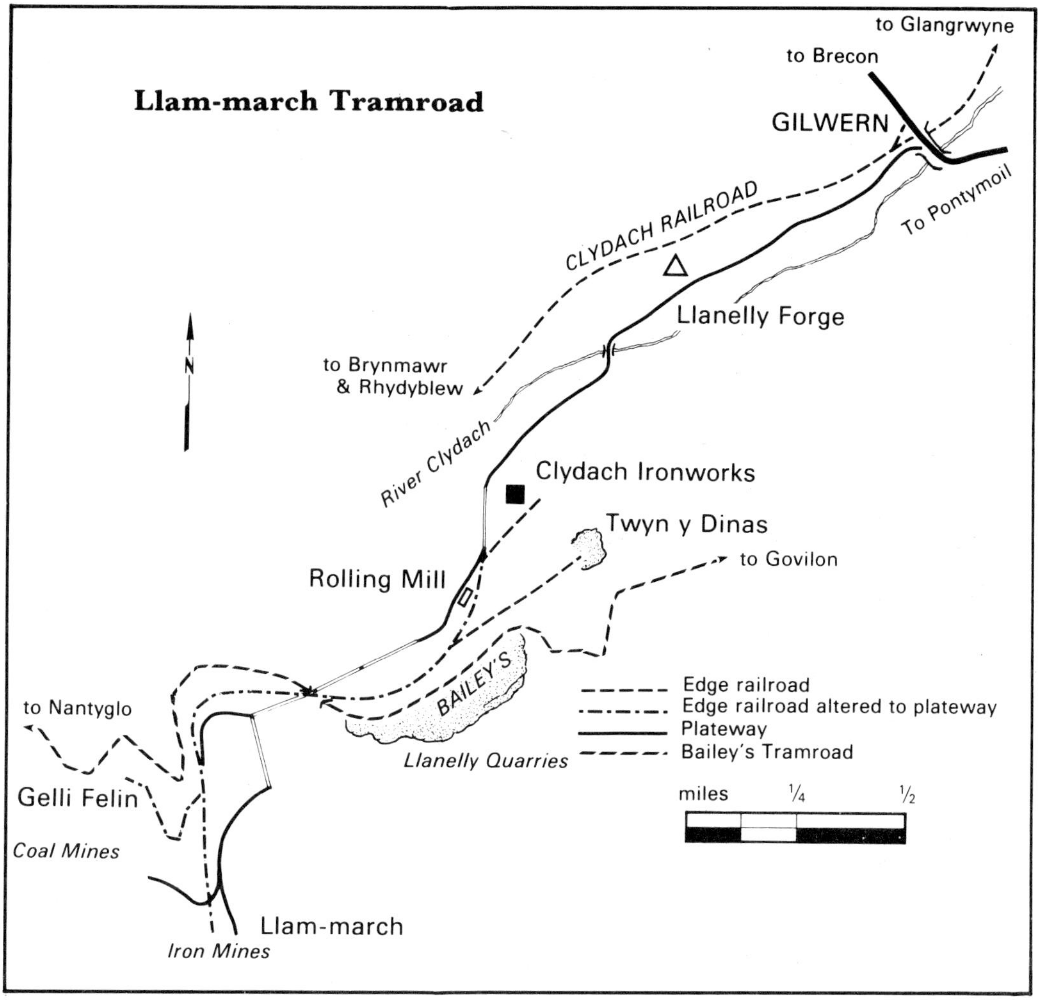

On 17 October 1793, the B&A assembly received two applications for railroads. One from Edward Kendall for a line from the Gelli Felen collieries he was intending to open, to run on the south of the valley, over Clydach Furnace bank to join the B&A's Clydach Rail Road, then under construction. The other was from Frere, Cooke & Kendall for a line from Lammerch Vach (Llam-march Fach) to "the furnace now building" and thence to the site of the intended canal, in all some two and a half miles. It is with the second that this section is concerned.

The B&A passed the matter to the committee, and on 28 May 1794, it decided that the company should construct the requested line from "... Llammerch Coal and Mine Works* to the new furnace now erecting at Clydach by Messrs. Frere, Cooke & Co. ... ", the iron company to pay a rental of eight per cent a year of the cost of construction, and to be responsible for all repairs during their tenure of the furnace. Should any other party carry on the line, they were to

* Mine = iron-ore.

pay a proportional amount of the repair costs, and the B&A reserved the right to take the line into their own management and to charge tonnages, should any further furnaces be erected. For their part, Frere, Cooke & Co. were to use no other means of carrying their raw materials.

Section 96 of the B&A Act specifically states that communications constructed under that clause should be " . . . to or from the said Canal or Rail Ways hereinbefore particularly described . . . " As Frere, Cooke & Co's. line was to terminate at the furnace, one and a quarter miles from the canal and eight chains away from the Clydach Rail Road, it appears that the clause was a little strained in its application!

John Dadford, the canal engineer, completed the survey of the new railroad by the end of June 1794, when the committee ordered work to be commenced immediately. Progress was quite rapid, and on 29 June 1795, Thomas Cooke suggested that the payment of rent should commence from 15 June. The accounts were not submitted for consideration until 31 August, however, and it was not until 14 October that all the figures had been checked, and the total expenditure agreed at £2,110. Orders were then given that an agreement with Frere, Cooke & Co. should be prepared, incorporating the terms previously stated, and including that the rental should be £168 16s. annually, commencing on 15 June 1795. It is possible that the railroad may have been in use a short while before this date, as at the beginning of June Frere asked the committee for permission to erect a weighing machine at the point where the line entered the works (SO.2296 1294). The offer of a machine that the B&A had purchased at cost price, £55 2s. 6d., was accepted on the understanding that should the canal company take over the management of the railroad they would buy it back from them.

The exact date on which the Clydach Ironworks came into blast has not been ascertained, but it must have been before May 1797, when the B&A ordered rails from them for the branch into Llanelly Wharf from the Clydach Rail Road.*

By late 1797 Edward Frere could see that it would be some years before the hoped for through communication with Newport would be ready and it was essential that some means should be found to export the produce of the Clydach Works. Accordingly, he approached the Monmouthshire Canal committee with the proposal that he should be permitted to make a stone road from the nearest point on the Llam-march Railroad at Waen Llapria (SO.2221 1221) to Blaenavon, where iron could be put onto the Blaenavon Railroad for carriage to Pontnewynydd, and then onto the canal to Newport. Such roads could of course be made under the eight mile clause of the MCC Act of 1792.

The MCC agreed, and at their committee meeting on 21 November 1797 arranged with Frere that he should make the road on condition that not more than one quarter of his annual tonnage for carriage over it should be spent on its formation, the arrangement to hold good for one year.[5] The agreement was renewed for a further year on 29 April 1799, and was probably renewed from year to year, although it is not mentioned in the MCC minutes. At their meeting on 2 May 1804, the MCC committee agreed to allow the Clydach Iron Co. £150 towards the cost of repairs to the road " . . . for which they agree to use the Blaenavon [Rail] Road 'til the Abergavenny and Breconshire Canal joins the Monmouthshire . . . " Such was the penalty of having a works at a place remote from existing transport!

The railroad made under the auspices of the B&A did not satisfy the needs of the Clydach Co. for long, and on 17 October 1798 the B&A committee minuted that:

> "Messrs. Frere, Cooke & Kendalls having represented that the Lammerch Rail Road is not equal to

* This is as stated in the minute book, but it is probable that there was, at the time, only one wharf at Gilwern—the one later known as Clydach Wharf (see page 68).

> the business of Two ffurnaces and having proposed to alter the same at their own Expence into a Double Dram Road, Resolved that they may be at Liberty so to do and that a proper Contract be entered into with them for that purpose but that with the Proviso that if this Company shall think proper to take Tonnage on the said Rail Road instead of rental that Messrs. Frere & Co. are to be refunded the Expence of such alteration."

It has not come to light in what precise manner the railroad was unsatisfactory, but it is probable that, after the construction of the stone road to Blaenavon, it was found that the railroad waggons were too heavy when carrying loads upwards on a 1:9 gradient for seven furlongs. It is interesting to note, however, that this proposal was put forward *before* Benjamin Outram had been asked to report on the B&A railroads, which was not until June 1799.[6]

A draft agreement was prepared in 1800,[7] but it was not until 29 April 1802 that the assembly ordered the committee to read it and bring it forward to be sealed at the next assembly.

Construction seems to have been somewhat delayed, and it is not until April 1804 that news of its progress is forthcoming, when Frere wrote to John Powell:

> ". . . our Coal Road was going on glibly last week I was over it on Saturday and intended seeing it again before I met you I think ten Days would enable the Trams to pass along it and I do not fear of its being in regular working in two or three more . . . "[8]

but it is not known when it was finally completed.

The B&A was opened from Gilwern to Brecon in December 1800,[9] and in March 1802, a special assembly was held to consider the expediency of completing it to Pontymoil to connect with the MCC. At the regular assembly on 29 April, it was decided to apply to Parliament for powers to raise the necessary additional capital. Some of the Breconshire proprietors were content that the canal should remain in its present state, the supply of cheap coal to the county town being their sole concern. Frere, however, was all in favour of the extension, as he could see that there was a danger that the MCC would acquire a monopoly of the trade from the major ironworks. Once this happened, the B&A, to say nothing of the Clydach Ironworks, would be left on a limb. As he put it, in a letter to John Powell written before the special assembly:

> ". . . but they are busy bustling people & if we sit with our eyes shut & our hands in our bosoms until they have provided all the custom they want to themselves we may then make a fish pond with little danger of the fish being disturbed but to call it a canal will be quite a joke . . . "[10]

Frere also realised that once the B&A was connected to the MCC it would be essential to his Clydach Ironworks that it should have an easy route to the canal to enable it to take advantage of clause 108 of the B&A Act, and on 20 October 1803 the iron company applied to the assembly for a lease for ninety-nine years on the site of the feeder from the river Clydach to the canal in order to extend the tramroad, and a site close to the aqueduct at Gilwern on which to build a warehouse. The committee considered the proposal, but reported that it "does not appear prudent to discontinue the Clydach feeder,"[11] but it is probable that the true reason for the rejection was that, at the time, Gilwern was the terminus of the canal, and a further tramroad connection would add to the existing congestion.

Six years later, by which time the canal had been extended to Llanfoist, and the demand on wharf space at Gilwern had eased, the Clydach Co. made a similar application, but this time suggested that the tramroad should be laid on the side of the feeder. Scenting an opportunity to pass off some of the maintenance charges, the committee acquiesced on condition that the iron company undertook to make all repairs to the banks of the feeder, and an application to build a warehouse adjacent to the aqueduct was granted at £1 1s. a year, and an undertaking to maintain the waste weir into the Clydach at this point.[12] On 26 April 1810 Frere, Cooke & Powell were given permission to construct a wharf of one boat's length to serve the warehouse.

It appears that the reconstruction of the railroad as a plateway was unsatisfactory, and on 24 April 1806, the assembly was asked to approve a tramroad on an entirely new line. The B&A approved and the line was sanctioned subject to the entire expense being borne by the iron company,

> " . . . and that a proper contract be entered into . . . for the purpose, with the proviso that if this Company should think proper to take the tonnage of the said Road instead of Eight per cent of their expenditure, Messrs. Frere, Cooke & Powells are to be refunded the expence of such alteration now proposed".

The new line as constructed was very much on the route suggested by Outram for a revised Clydach Rail Road;[13] the portion of the existing tramroad adjacent to the works to Daren Felen, on the line of the original railroad made by the B&A, was abandoned, and a series of inclines falling from Daren Felen to Glan Sychnant, and running level thence to the coalyard and the charging bank of the furnaces was substituted.

The new line was completed by 17 October 1811, when Frere, Cooke & Powell requested the assembly to appoint someone to examine their accounts to ascertain the sum to be paid by the B&A should they purchase the line. No action was taken until 21 October 1813, when on Frere's pressing the matter, John Hodgkinson was appointed to make a survey and report, apparently as a prelude to the canal company taking the line over.

No record has been found of the cost of the conversion, and it is impossible to calculate this from the published accounts, which show one figure only for the rents of tramroads, railroads and houses. It is, however, certain that the tramroad was purchased by the B&A, as on 24 April 1834 an application was made to them for a reduction in the rent, and in 1861, after the Clydach Iron Co. had failed, when the B&A sought repossession of the line, their solicitor reported that he had found that there was a lease " . . . of the Road as a Rail Road to Messrs. Frere, Cooke & Co., for a term of years as yet unexpired . . . "[14]

Edward Frere retired from the canal committee on 29 April 1813. He had always looked at canal affairs from the viewpoint of the welfare of the ironworks, and with their transport in a reasonably satisfactory state, he evidently considered the time was ripe for him to concentrate his activities on the works. In any case the works *vis-à-vis* the B&A was adequately represented by Walter Powell of Brecon, who in addition to being his partner in the works, was solicitor to the canal company.

A considerable amount of metal was exchanged between the various ironworks as a means of getting rid of surplus iron or making good deficiencies in production. Clydach was badly placed in respect of this aspect of the trade, being somewhat outside the area normally associated with the smelting side of the industry. To reach Blaenavon it was necessary to resort to road haulage, and to reach any of the works between Nantyglo and Tredegar involved carriage to Gilwern on their own plateway, transhipment to the B&A's Clydach Rail Road, and carriage back up the valley passing the works *en route* on the other side of the river, but about seventy-five feet higher.

To obviate this difficulty and unnecessary expense, the iron company applied to the assembly of 16 October 1817 for a "Rail Way or Stone Road" from the Clydach Rail Road to the works, stating that they would make it themselves under the eight mile clause should the B&A decline to do so. A week later, the committee turned down the request, but it seems that the iron company made the road, possibly as a stone road, as there is a small iron bridge over the river adjacent to the works (SO.2291 1332), and a road thence to the line of the Clydach Rail Road at SO.2297 1362 was severed in recent years by the Heads of the Valleys Road, although a part remains above the main road. The distance covered by this from the bridge was 370 yards, which corresponds closely with the "Three Hundred and Fifty Yards or thereabouts" which the B&A was asked to make. Cast into the beams of the bridge is the date "1824".

J. & C. Bailey of Nantyglo applied to the B&A in October 1820 for the construction of a tramroad from their works to the canal at Govilon, to which Edward Frere lodged a written objection. The proposal was rejected at the assembly in April 1821 on the advice of the committee, but the Baileys were given permission to make it themselves. Bailey's Tramroad was completed by December 1821, and objections to it were heard and damages assessed by Commissioners appointed under the B&A Act on 30 June 1824. Frere, Cooke & Powell were given small damages in respect of interference with the upper of their inclines on the Llam-march Tramroad at SO.2221 1248, where their embankment had been burrowed through.[15]

In 1844, Samuel Lewis described the Clydach Works as consisting of four furnaces, with a rolling-mill driven by a forty feet water-wheel, and states that:

> "An immense quantity of coal is also raised here [as well as iron ore] by the Clydach Iron Company for the supply of their extensive works; it is all worked by levels, brought down the mountain steeps by inclined planes, and conveyed in trams by the railroad belonging to the canal company, either for the supply of the ironworks, or to the Brecknock and Abergavenny Canal . . . "[16]

The method of working the inclines is given in a manuscript in the possession of the Abergavenny Museum written by Thomas Jordan in 1909, in which he states:

> "The materials . . . for the use of the ironworks, were brought down from the hillsides by a series of inclines—each about three hundred yards long, forming a zig-zag line down the mountainside. From the top of each incline to the foot of the next, was a level piece of ground, forming a landing-place, where two men were stationed to pass the trams from one incline to the other. The inclines were self-acting, having the weight of the full trams going down drawing the empty ones up, and were controlled and regulated in their motion by a brake on a wheel over which an endless chain worked, and to which the trams were attached . . . "[17]

This first-hand account tallies well with the specification for the inclines on the Llangattock Tramroad, which laid down that there should be:

> " . . . two horizontal wheels and an endless chain of the same substance as that made use of by Messrs. Frere, Cooke & Powells at Clydach Ironworks . . . "[18]

What might possibly have been the salvation of the Clydach Iron Co., the Merthyr, Tredegar & Abergavenny Railway which planned to lay a mile-long siding to the works in 1858, was too late, and before it was built Nemesis had overtaken the iron company. On 4 July 1861, it was reported to the B&A committee that the Clydach Co. owed £1,650 in respect of tonnages, and at the end of the month the solicitor reported that he had obtained judgement against them for this amount, but that he had not yet served it. In August he informed them that he had lodged the writ with the Sheriff as a result of information that had come his way, and in October the B&A was in possession of the works. A consequence of this large amount owing to them was that the B&A was unable to declare a dividend in October 1862.

Of the three trustees in bankruptcy appointed to clear up the affairs of the Clydach Iron Co. only one, John Jayne, the former 'truck-master' at Nantyglo, appears to have been active, possibly due to the influence of Crawshay Bailey to whom the property reverted in his capacity as successor to the Kendalls in the Beaufort Works.

In its early days it is probable that the Clydach Works obtained its limestone for fluxing from the very old quarries at Twyn y Dinas (SO.2285 1255), from which a plateway and incline have been traced. Indications on the ground are that this connected with the Llam-march Tramroad at the foot of the upper incline at Daren Felen. This source must have been disused by 1821, as

Smart's Bridge, Clydach Ironworks.

part of its line was made use of by the Bailey brothers for their tramroad to Govilon. After this date, it is likely that stone was obtained from part of the site of the later Llanelly Quarries (SO.223 124) and taken to the works from the same point on the inclines.

The Llanelly Quarries, as such, were opened by John Jayne under a lease from the Duke of Beaufort in 1863, which granted him the right to make a new tramroad and incline to the ironworks site which might be used by any future tenant.[19] The plan accompanying this lease is marked in pencil "This Tramway not made", and the old inclines must have been kept in use.

In 1864, the site of the furnaces and rolling-mill were sold to the New Clydach Sheet & Bar Iron Co. Ltd., another grossly undercapitalised concern, whose working capital was derived largely from the issue of debentures. By the time the Liquidator was appointed in 1867, Jayne had contrived that most of the debentures were in his name, so that when the works site was sold by auction in October, he was able to purchase it for a mere £960, the remainder of the purchase money being accounted for by his debenture holding.[20]

Jayne died in January 1873, and his executors let the works, now valued at £220,000, to the Brynmawr Coal & Iron Co. Ltd. From the start this company was short of funds, and immediately issued debentures to the value of the works, and proceeded over the next few years to raise money by borrowing from their bankers; by accommodation bills of exchange; and by second and third mortgages. The end came in November 1877, when a receiver and manager was appointed. The executors, as their principal before them, had by this time acquired all the debentures, and by the time the company's affairs were wound up, and attempts made to sell the property had failed, the iron company's bankers were glad to accept £5,500 in full settlement of a debt of £12,117. In August 1881, the property of the company was put into the name of an Abergavenny solicitor to administer in trust for the beneficiaries of John Jayne's will.[21]

It is not known what use was made of the works by the two later holders, but it would have been essential for them to have kept the Llam-march Tramroad in use to obtain their raw materials. A plan of the whole estate involved, accompanying the assignment of the lease to Jayne in May 1869, shows that by that time a connection had been made by the MTA to the foot of the upper incline at SO.2224 1249 from a siding in Llanelly Quarry; this is also shown on the 1964 issue of the 6″ O.S. Research in London & North Western Railway records may reveal the extent of the use of this siding.

Between the ironworks site and the canal, a forge had been established about 1690, which was taken over by the Clydach Iron Co. in 1800 for use as a tinplate mill. This was served by the tramroad to the canal authorised in 1809. In 1864, this works was disposed of as a separate entity to John Miles of Pontymister, and a wayleave granted on the tramroad between this point (SO.235 139) and the canal.[22] Presumably this line was used by Miles until the works was closed in 1884,[23] when probably the whole tramroad went out of use.

Route
Railroad 1795 and Plateway 1804

From the Clydach Iron Co's. coke yard at SO.229 131, the line ran uphill in a south-westerly direction, on the line of the minor road to Daren Felen, crossing the line of the later MTA at SO.2223 1249, where the railway company later made a level crossing. It continued along the road to SO.2185 1228, where it divided, one line remaining on the site of the road to cross the stream at SO.2194 1199 and following the road to the coal workings; the site of these has been lost in later operations, but they are stated by Lloyd to have been to the east of the Rhassa (i.e. Nant Gam). The other branch climbed steadily from SO.2185 1228 to the iron mines on Llanelly Hill.

Plateway to Canal 1809

This possibly the most difficult part of the Clydach Works' tramroads to trace, due to the many alterations of recent years.

The line started at the tapping floor of the furnaces at SO.229 132, running in front of the furnaces it crossed the Nant Dyar at SO.2300 1333. A new road was made from this point to SO.2332 1375 as part of the alterations necessary when the Heads of the Valleys Road was constructed. This was formerly a narrow path, marked out with stone blocks, and was also used as a 'slag-road', as is witnessed by the iron slag at SO.231 134.

The Clydach river was crossed at SO.2318 1374, where the bridge still stands, although well concealed. The Heads of the Valleys Road now intervenes, and the tramroad cannot be found with any certainty until SO.2335 1384 is reached. Passing Forge Row, the line keeps to the north of the river, but at SO.2354 1395 a branch led to the other side, apparently terminating at the *Forge Hammer* public house (SO.2375 1405). Still on the north of the river, the main line continued, to cross the railroad to Glangrwyne at SO.2434 1446, and entered the wharf site at SO.2434 1446 to the east of the present dock; it then turned south on the off-bank of the canal to the iron company's wharf.

Plateway and Inclines 1811

It is not known if the incline from the tapping floor, SO.229 132, to the level of the charging bank, SO.2289 1307, was constructed as part of this alteration, or if it had been made previously. It was from the latter point, however, that the final alterations were made in 1811.

From the top of the charging bank the tramroad led to the west of the original railroad at SO.2286 1301, and ran on the line of the present road through the small housing estate, the stone wall on the eastern side of the road representing part of the site of the rolling mill.

After passing the estate, the line passed through a shallow cutting and turned to run in a generally south-westerly direction, passing about forty yards to the north of a house, Glan Sychnant, and passing over the water course to the works at SO.2251 1267. The water-course was diverted and culverted to pass under the foot of the incline at right-angles.

From SO.2251 1267, the line rose by three inclines each of about 190 to 200 yards in length, climbing from the 550 feet contour to just over the 1,000 foot line. The first stage finished at SO.2240 1255, the second at SO.2224 1249, where the line of the original railroad was crossed, and the third at SO.2204 1242, after crossing Bailey's tramroad at SO.2220 1247.

From SO.2204 1242, the tramroad was carried on the level by a side-cut in the hillside, to join the original railroad at SO.2185 1220.

A new means of reaching Llanelly Hill was also brought in when the inclines were introduced. From the head of the third stage (SO.2204 1242) a further incline was made, rising to SO.2221 1221 at Waen Llapria, where the line divided, the eastern branch leading to workings on the north of Llanelly Hill at SO.229 120, and the western branch swinging round to the south on the line of the existing road to Llam-march. A branch from this line turned off to the west at SO.2209 1195, to cross the stream at SO.2193 1178, a few yards to the north of the present bridge, where traces can be seen of the original abutments. A plate, three feet long, was recovered from the western abutment in 1974, but from this point all traces of the tramroad are lost under recent workings.

REFERENCES:

1. GRO. D.591/42.9.
2. ibid.
3. GRO. D.591/23.46.
4. Lloyd, J. *Early History of the Old South Wales Iron Works, 1760–1840.* 1906.
5. PRO. MCC committee minutes, 27 November 1797.
6. NLOW. Maybery 383.
7. ibid. 776.
8. ibid. 3188.
9. Hadfield, Charles. *Canals of South Wales & the Border.* D&C. 1967.
10. NLOW. Maybery 772.
11. PRO. B&A committee minutes, 21 November 1803.
12. ibid. 4 August 1809.
13. NLOW. Maybery 383.
14. PRO. B&A committee minutes, 17 October 1861.
15. GRO. D.591/39.100, 39.91, and 39.85.
16. Lewis, Samuel. *Topographical Dictionary of Wales.* 1844 Edn. Article, *Llanelly.*
17. Jordan, Thomas. *My reminiscences of the Old Clydach Iron Works.* 1909. Unpublished MS held in Abergavenny & District Museum. The passage quoted is reproduced by kind permission of the Curator. Thomas Jordan was a grandson of Robert Smith, mining engineer to the Clydach Works.
18. NLOW. Maybery 964–5.
19. GRO. D.591/3.241.
20. ibid. D.591/3.246.
21. ibid. D.591/3.286.
22. ibid. D.591/19.274.
23. Rees, D. Morgan. *Mines, Mills and Furnaces.* H.M.S.O. 1969.

9

Llangattock Tramroad

The importance of lime to the farmers of the Usk valley in the nineteenth century can be judged by the frequency with which limekilns occur along the line of the Brecknock & Abergavenny Canal and its attendant tramroads. In the main, the older kilns were close to the quarries, but with the advent of better transport provided by the canal, the later ones were built on sites either on or below the canal banks.

In the final quarter of the eighteenth and the first of the nineteenth centuries, the Duke of Beaufort let large areas of his extensive property in south Breconshire for exploitation of the minerals, and at the canal assembly on 21 October 1813, a request was received from His Grace for the canal company to construct a tramroad from his quarries at Darrenykilley (Daren y Cilau) to the canal at Llangattock Wharf. The matter was referred to the committee who instructed John Hodgkinson, the engineer, to survey the proposed line, and to present his plans and estimates at their next quarterly meeting and to provide an estimate of the probable cost of lime from this source at Brecon. He reported to the meeting on 17 January 1814, and a resolution was passed that the scheme appeared highly beneficial " . . . to the Publick and the Company . . . ", and that the estimate should be submitted to the Duke for his consideration.

The Brecknock Boat Company—a partnership consisting of John Powell, the B&A solicitor; their banker and treasurer, Jeffreys Wilkins; and two substantial shareholders, John Lloyd and John Peirce—had been formed in 1796 to promote a trade in coal and lime on the canal. They already held several leases of quarries and collieries in the Clydach valley, and saw in the Duke's suggestion a possibility of increasing their business, and the February committee meeting received a letter from them offering to construct " . . . the intended Tram Way . . . with a proper convenience for delivering Lime Stone from the Rock to the said Road including fences, lands, lime kilns and all the requisite conveniences for a sum not exceeding 2,000 pounds and do agree to rent the same for a term of 21 years at a clear Five per Cent of the Sum laid out . . . " The committee promptly resolved to recommend acceptance of the offer to the assembly.[1]

The assembly endorsed their recommendation on 28 April 1814, and authorised them to borrow £2,000 for the construction, and to employ David Davies of Crickhowell to survey and value the necessary land. The decision was notified to the Duke, and on 21 June his agent reported that he was agreeable to let the quarry to the canal company for twenty-one years at a galeage (royalty) of 1d. per ton. A letter from Thomas Ellwood, the BBCo's. mineral agent, dated 3 June 1815, quoting a break-down of the cost of stone at the wharf, shows that the B&A

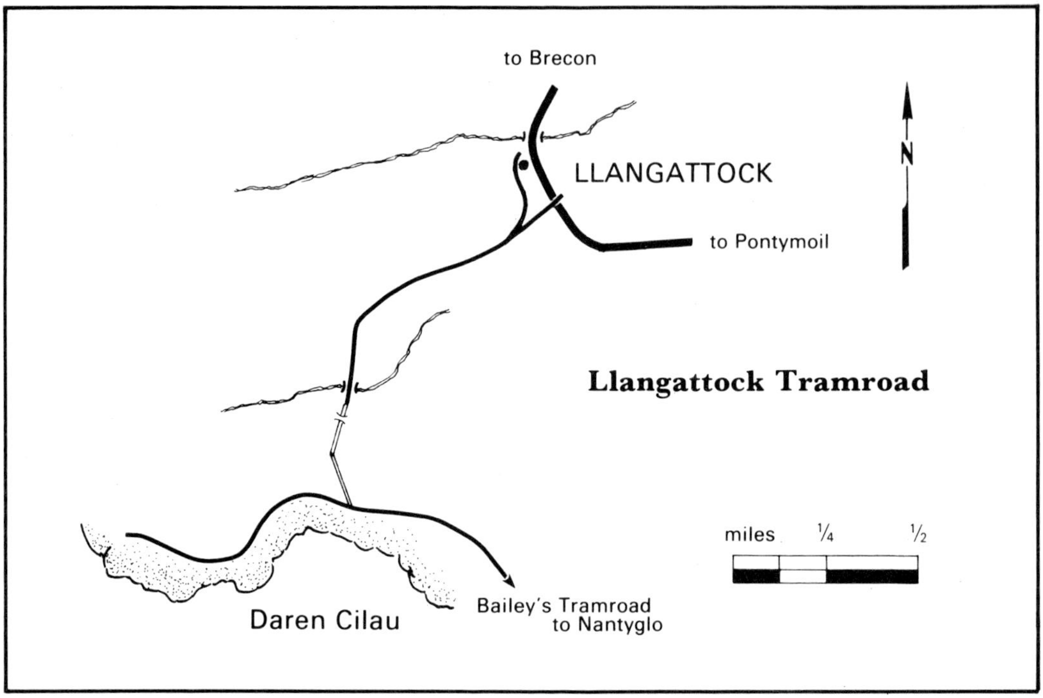

permitted the BBCo. to extract stone at 1d. per ton, thus making no profit from their lease of the quarry, relying solely on the tonnages to be received from the carriage of the lime and stone on the canal to receive any benefit.[2]

By September 1814, the subscription list, headed by the Duke, was filled for the authorised sum, so the committee gave instructions for the land to be purchased,[3] and the solicitor was ordered to prepare a contract for the construction of the tramroad with the BBCo., their liability to pay interest to commence as soon as they received any payment.

A month later the route had been marked out, but in a letter to B. A. Griffiths, the canal Agent, dated 14 October, Ellwood informed him that he had found a way of by-passing an orchard through which Hodgkinson's suggested line passed, which made little change in the length or the gradient of the line. It is apparent from this letter that the BBCo. did not intend to make the tramroad by direct labour, but " . . . should you see the contractors I shall thank you to inform them that on Tuesday next, the 11th I shall be ready to point out the different line to them . . . " He also asked that Griffiths should attend to give his opinion.[4]

The contract with the BBCo. was comprehensive, and laid down a strict specification of the manner in which the line was to be made. The road-bed was to be eighteen feet wide (wide enough for a double track), and bounded by walls, with ditches two feet six inches deep, and a " . . . three feet wattled hedge to be properly quickened with hawthorn setts of not less than seven years growth . . . " An inclined plane was to be constructed to consist of " . . . two horizontal wheels and an endless chain of the same substance as that made use of by Messrs. Frere, Cooke and Powells at Clydach Iron Works and to be complete in every respect . . . " The stone blocks were to weigh a minimum of 168 pounds, and the plates were to be three feet long and to weigh at least fifty pounds. The BBCo. had to maintain the line for one year after completion, and payments were to be made by the B&A periodically on engineer's certificates of work done.[5]

By January 1815, work was well under way, and the committee authorised a payment of £872 16s. 4d. on account;[6] and it would seem that the incline was ready for use by June, when Ellwood reported the somewhat disastrous first attempts to use it in the following terms:

> "As for the state of the Llangattock Road I am sorry to have to say that we went very ill by defects in the machienery chains &c. caused by a breakige of some of the trams but at present it begins to work very well and I have no doubt that the Incline Plain will work very well. The principal Injury done was in breaking Tram Wheels there is but one Tram left but what can be imidately fitted up Eight was the number injured Two of them since fitted up and only waiting for Wheels when we shall have again all on Works Hoping that there will be no more such accidents but it is mostly the Case in all new Erections not to go so well at the Commencement every Thing being strange . . . "[7]

The whole tramroad was probably ready for use at the end of September 1815, but the lease to the BBCo. was not signed until December and was back dated to 29 September 1814, the date of the contract for construction.[8]

In January 1816, the BBCo. applied to the B&A for the advance of a further sum of money to ". . . complete the Llangattock Railway . . . " and to render the "shoot" at Daren y Cilau more efficient. The committee considered £92 15s. sufficient to cover this operation and approved the advance on condition that the BBCo. should pay five per cent per annum on this sum in addition to the rent already payable for the tramroad.[9] In August 1816, when the accounts were finalised, it was found that the total cost of the tramroad and ancillary works had been £3,000 making the annual rental £150.[10]

Even two years later, in August 1818, not all payments for subscriptions had been received, and Walter Wilkins who had lent the B&A £1,000 on the security of a promissory note which was then due for payment, had his £200 subscription to the tramroad deducted from the money he received. That he was not the only defaulter is borne out by the instructions given to Griffiths that he should take steps forthwith to collect all outstanding sums and hand them to the treasurer, who should then transfer the balance of the tramroad account to the general account.[11]

The "shoot" mentioned in the BBCo's. application of January 1816 seems to have been a source of recurring trouble, and in August 1818, Ellwood reported to his principals that "The slip at Llangattock certainly threw us back in getting a regular supply of Lime Stone . . . "[12] and in January 1819 he reported:

> ". . . I have further to observe that Llangattock concern is going on as well as Circumstances will alow but I am sorry to add that the shout [*sic*] for rolling down the Stones doesn't answer the expectations although I was Myself very Sanguin of its utility at the Commencement—the Distance the Stones has to roll down breaks the Banks and loosens the Earth so much that when Rain comes it washes down such Quantitys of earth so as to make it detrimental to filling the Stones and causes Extra Expence in clearing it away— should it be found that the Lime Trade should increase . . . I should then recommend the adoption of Inclined Plains to the Rock which strikes me ought to be looked at by the Canal Co. . . . "[13]

The contract of September 1814 called for the construction of one incline only, but as there are now the remains of two on the line of the tramroad, it must be assumed that the B&A took note of Ellwood's suggestion; there is, however no mention in their minutes of its construction. The fall of the upper incline is approximately 300 feet, and the havoc caused by the stone rolling from this height for some 110 yards can well be imagined.

Ellwood also recommended that a residence for the "accountant" (probably the wharfinger) should be built at Llangattock so that he should always be on hand. In January 1820, the canal

Llangattock Wharf and kilns. The Llangattock Tramroad terminated on top of the kilns.

committee authorised the building provided that it should not cost more than £90. The house was completed in July 1821, and the £90, plus £60 spent on improvements to the wharf added to the sum on which the BBCo. was paying five per cent.

When it came to paying for the land used, the B&A was not so prompt, one former landowner not being paid until 1822, and in July 1832, John Jones, who was now a partner in the firm of solicitors watching over the company's affairs, had to request payment of £300 in order that he might wind-up an estate he was administering.[15]

In January 1827, the B&A gave the BBCo. permission to extend the tramroad westward to fresh quarries, but instead of the usual *carte blanche* given to requests from this source, it was stipulated that the B&A should lend them the rails, to be returned when the operation was completed.[16] It is not known why the usual free and easy response was lacking on this occasion, but it is possible that the committee had heard of J. & C. Bailey's proposals to extend their coal business in the Brecon direction.

The B&A received a request from the Baileys on 24 April 1828 for a tramroad from Llangattock to Nantyglo. The committee considered the proposal but decided on 24 July that it would not be in the company's interests to construct the line, but that no impediment should be put in the way of the Baileys making it for themselves. The Baileys set to work immediately, and probably completed their tramroad from Nantyglo to the head of the upper incline by mid-1831.

Ten months later the B&A was notified by the BBCo. that they had assigned the lease of the Llangattock Tramroad and the kilns at Llangattock to J. & C. Bailey, "... who have announced

Llangattock Tramroad: Head of upper incline.

their intention of uniting their new road to the Llangattock Rail Way at the top of the present incline plain . . . "[17] The Baileys also offered to purchase the incline at the cost of constructing a new one, subject to valuation. No action seems to have been taken until January 1831, when in response to an application from the Baileys, the B&A appointed Ellwood to make the valuation and to report to the next assembly. The sale was authorised on 28 April 1831, but the purchase price is not recorded in the minutes. Ownership of the tramroad below the inclines and the kilns was retained by the B&A, and when the BBCo's. lease expired at Michaelmas 1836, the tramroad was let to the Baileys at £105 a year, and the kilns were re-let to the BBCo. at £25 per annum.

The BBCo. continued to use the tramroad and the quarries, presumably paying tonnages to the Baileys for the use of the incline and the tramroad. In fact, their trade seems to have increased, and in April 1844 they applied for and were granted permission to construct another kiln at Llangattock on the usual terms of five per cent of the cost of construction as an annual rental. At that time, Llangattock must have been quite a busy place, with the Bailey's coal wharf, and the Boat Company's trade in coal, lime and limestone. Unfortunately, no records have been found of the working of the tramroad under J. & C. Bailey.

By 1861, trade on the canal was declining rapidly. The Monmouthshire Railway & Canal Co. had completed the alteration of its tramroads to locomotive hauled railways in the previous year, and had provided a cheaper and quicker route to Newport for the ironworks which had formerly taken advantage of the cheaper rates on the B&A. In this year the B&A was glad to renew the lease of Llangattock Wharf and kilns to the BBCo. at £15 a year.[18]

The tramroad was sold to the MR&CC with the rest of the B&A property under the Act of 1865 for the amalgamation of the two canal companies.[19] The Act mentions it specifically as a "... Railway extending from the Canal ... to Messrs. Bailey's incline Plane ... " As the BBCo. ceased to trade about the same time, and J. & C. Bailey had adequate direct communications by rail to Newport from both Nantyglo and Beaufort Works, it is unlikely that much use was made of the Llangattock Tramroad after 1865, although the land from the foot of the inclines to the canal remained in the possession of the MR&CC until that company was acquired by the Great Western Railway in 1880, as is witnessed by the numerous GWR boundary markers to be found.

Route

The tramroad commenced at the top of the kilns on the canal bank at SO.2065 1727, whence it proceeded to the west of Wharf Cottage to cross the road from Llangattock to Beaufort at SO.2059 1721. Continuing south-easterly and curving round to the south-west it joined the by-road from Gilwern at SO.2066 1705, and ran alongside this to SO.2058 1692, where it passed through the gate. There are a number of blocks in evidence on this section.

Llangattock Tramroad and quarries.

Through the gate, there is no possibilty of missing the line which climbed steadily to cross Nant Onneu at SO.1999 1639. From this point to the foot of the incline the track is obstructed by brush-wood, but it is possible to pass on the outside of the hedge.

The incline commenced at SO.1999 1634, and climbed to SO.1986 1610, passing to the west of a coniferous plantation. At the top there is a circular, bracken filled depression marking the site of the horizontal brake wheel, and close by are the remains of the brakesman's shelter. Formerly this incline was crossed by an occupation bridge at SO.1997 1631, but this has been removed in recent years.

The second stage of the incline rose to SO.2004 1591 where there remains the retaining wall for the final twenty yards. Here too, there is a circular depression, some ten feet in diameter, where the brake wheel was situated. Where the chain passed over the lip of the incline there are several stones cut into a groove by the passage of the chain. To the east of the depression there is a small, square, stone.lined hole through which the brake-rod passed into the adjacent brakesman's shelter, set slightly to the east, to enable the trams to be watched passing up and down the incline.

From the head of the upper incline, the line ran westward on a ledge cut into the steep hillside, and can be followed to SO.1910 1574, passing the remains of a weigh-house at SO.1925 1568, at the entrance to a small subsidiary quarry. There are many stone blocks remaining on this section.

Bailey's coal wharf was established at SO.2075 1720, and to reach it the canal was crossed by a draw-bridge. From this, there is an embankment running south to join the main line of the tramroad at SO.2066 1705.

This plateway was to a gauge of three feet six inches (measured from back to back of the plates). and from SO.2059 1721 to SO.2058 1692 most blocks bear evidence of having been used with the plates spiked to them, with a plate length of three feet. Above the incline, and from SO.2058 1692 to the foot of the lower incline, most of the blocks bear the characteristic 'spade' mark indicative of the use of tie-bars. The latter form of track may have been laid during the tenure of the Baileys, who favoured this type of construction. Above the incline movement of the blocks makes it impossible to measure the length of the plates used with the tie-bars, but an average figure of forty-two inches was arrived at from numerous measurements taken.

Whilst the track-bed below the inclines is of ample width to accommodate double track, it is evident that in its later days only a single track was in use.

REFERENCES:

1. PRO. B&A committee minutes, 7 February 1814.
2. Brecknock Museum, Brecon. Maybery Collection, Vol. 2 fo. 111.
3. PRO. B&A committee minutes, 15 September 1814.
4. NLOW. Maybery 960.
5. NLOW. Maybery 964-5.
6. PRO. B&A committee minutes, 7 January 1815.
7. Brecknock Museum, Brecon. Maybery Collection, Vol. 2. fo. 111.
8. GRO. D.591/11.61.
9. PRO. B&A committee minutes, 22 January 1816.
10. ibid. 7 August 1816.
11. ibid. 11 January 1818.
12. NLOW. Maybery 1106.
13. ibid. 3287.
14. PRO. B&A committee minutes, 4 November 1820.
15. ibid. 16 October 1822, and 26 July 1823.
16. ibid. 25 January 1827.
17. ibid. 23 July 1829.
18. ibid. 17 October 1861.
19. 28-29 Vict. cclxxxi.

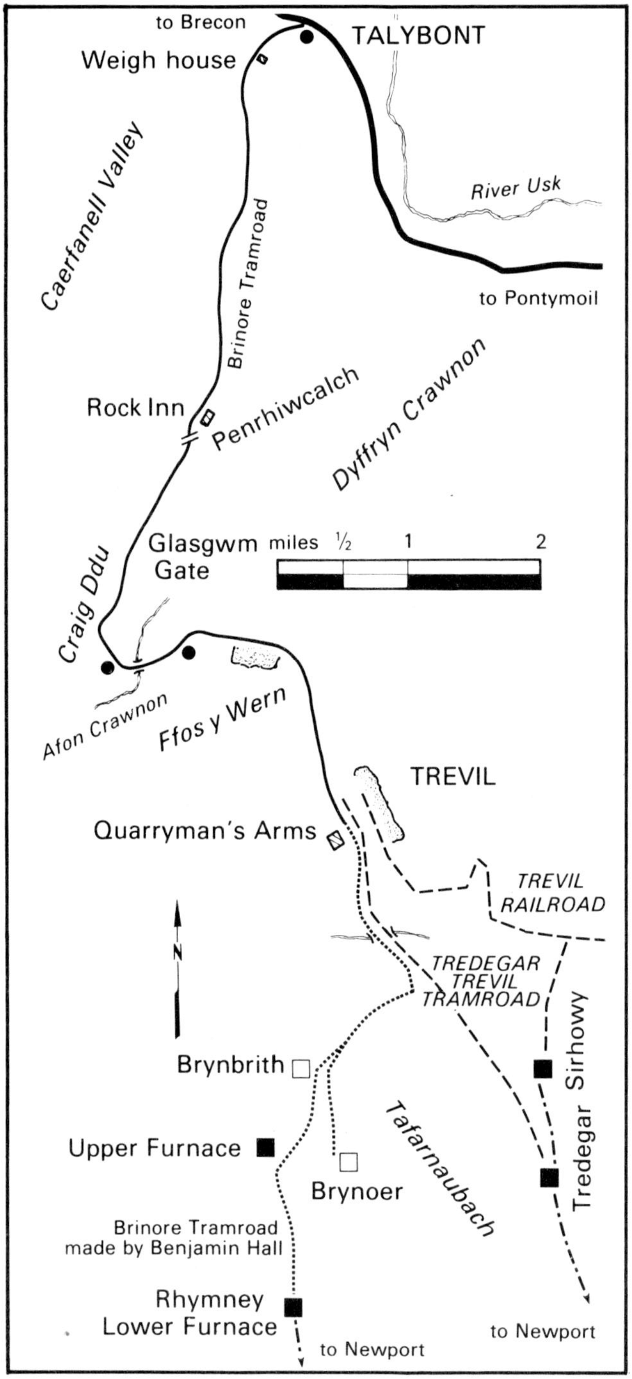

Brinore Tramroad
Adapted from 1st Edition Ordnance Survey

10

Brinore Tramroad

The Rhymney Valley, the most westerly of the valleys of the old county of Monmouth, was not sufficiently industrialised by the end of the eighteenth century to be worthy of the notice of the promoters of the Monmouthshire Canal, and it was not until 1823 that a scheme to provide direct communication with the port of Newport appeared in the shape of the Rumney Railway Co., usually known as the "Old Rumney" to distinguish it from the later Rhymney Railway.

At the beginning of 1800, a company largely financed by Bristol merchants erected a furnace at Llechryd, to the north of Rhymney, which was put into blast on 5 June 1801. In 1803, Richard Crawshay, Benjamin Hall and Watkin George of the famous Cyfarthfa Works, Merthyr Tydfil, purchased the Union Furnace as it was known, retaining two of the former partners as managers. George left the new partnership in 1805, and the remaining partners obtained a fresh lease for seventy-two years from the Duke of Beaufort in January 1807. In 1804, another furnace was constructed lower down the valley, in Rhymney. The two were then worked as one concern under the title of the Union Iron Co., the furnace at Llechryd being known as Upper Furnace, and the one in Rhymney as the Lower.[1]

Richard Crawshay died in 1810, leaving Hall in sole possession of the Union Co. under the terms of his will. With the cash element of his legacy, Hall purchased the Lordship of the Manor of Abercarne, in the Ebbw Valley, which had largely been developed industrially, and he must have seen for himself the benefit derived by that district from the presence of the MCC and its tramroads, and wished that his Union Works could have had similar facilities, instead of relying on pack animals. His opportunity came in 1812, when the Brecknock & Abergavenny Canal was opened throughout from Brecon to Pontymoil, where it met the MCC's Pontnewynydd line, providing a direct link to Newport. The B&A passed within ten miles of Rhymney, but on the other side of Llangynidr Mountain, and it was in that direction that his thought turned.

Accordingly he wrote to the B&A on 13 October 1812 requesting them to construct a tramroad from his quarries at Trevil to the canal at Talybont on Usk. After giving a detailed description of the route, he ended his letter, " . . . I should be obliged by an answer when you come to any determination on the subject as if you do not complete it I shall feel myself at liberty to do so under the powers of your Act".[2] His letter was considered at the assembly of 15 October, and referred to the committee, by whom it was discussed on 19 November, when orders were given to their Agent, B. A. Griffiths, to survey the route to ascertain that it came within the scope of the eight mile clause, and realising that Hall had been cunning in asking only for a tramroad from Trevil, and not from his other industrial interests which lay some distance beyond the eight mile limit, Griffiths was also instructed to find out from what distance any coal that might be carried on the proposed tramroad would have to come.

It was also realised that the placing on the canal of coal destined for Brecon at a point eleven and a half miles nearer that town than the existing point of shipment would lose the canal company a considerable amount in tonnages; and it is safe to surmise that the Brecknock Boat Co., whose partners were all considerable shareholders in the canal, pressed the committee to take strong action, as in addition to a reduction in their individual dividends, the partnership stood to lose its monopoly in the coal and lime trade on the canal. The same meeting in November resolved that a special assembly should be called to consider an application to Parliament for an amendment to the eight mile clause which would deny to owners of works the right to construct tramroads, but would retain the canal company's right to build them.

Griffiths reported to the December meeting, confirming that Trevil lay within eight miles of the canal.[3] He also produced an estimate of the comparative cost of coal at Talybont from the Clydach Valley and from Brynoer, showing the former to be 11s. 3d. per ton against 13s. per ton for the latter.[4] It appears that Griffiths' figures were somewhat inaccurate, as an estimate of the cost of coal *at Brecon* computed by John Llewellin, Hall's Agent, quotes Brynoer coal at 11s. 6d. per ton.[5]

The special assembly was held on 11 December 1812, and passed a resolution with no dissenting voices that the eight mile clause should be annulled. Wordy warfare ensued, culminating in a meeting of the 'County' of Brecon on 12 January 1813 to express their deep disapproval of the course the canal company was taking,[6] and evoking a letter from Samuel Homfray in answer to the B&A's fear of losses in their tonnages in which he offered to run the canal for twenty-one years " . . at the present income with an additional Hundred Pounds to it", provided the Act remained unaltered.

A further special assembly was called for 1 February by the opponents of the application to Parliament at which their proposed amendment to the previous resolution was defeated by 612 votes to 170.[8] The Bill came up for its second reading on 5 March, when on an amendment by Sir Charles Morgan, who had landed interests in the neighbourhood of Brecon, it was defeated.[9] Thus passed an attempt to perpetuate the monopoly of the Brecon coal trade for the BBCo., the principal beneficiaries being the solicitors for the two parties, who received £647 17s. from the B&A[10] and £604 11s. 11d. from the opposition.[11]

On 14 April 1813, a meeting was held in Brecon, chaired by Benjamin Hall ". . . to consider of the intended Railway from BRINORE to the . . . Canal at Talybont . . . " at which it was revealed that at least two surveys of the proposed tramroad had already been made by George Overton, a mining engineer, and John Llewellin. The meeting approved the ". . . shorter line of Road from Penrhiw Calch . . . ", and instructed Samuel Church, a Brecon solicitor, to make the necessary application to the B&A. It was resolved that should the canal company refuse to make the line, a subscription of £14,000 should be opened to make it.

Church attended the canal assembly on 29 April, stating that he was authorised to withdraw Hall's application. He then proceeded to make an application in his own name for two tramroads, one from his colliery at Brinbrith (SO.1130 0906), near Rhymney to Talybont, and the other from his quarry in Dyffryn Crawnon (SO.0973 1497), to the same place. It is not known why this procedure was adopted, but it may have been felt that in view of the slightly devious nature of Hall's application, another in his name would automatically be suspect.

The assembly of 21 October rejected the application, and Church published a notice in the *Cambrian* of 11 December stating that as the B&A had declined to make the tramroad, it was proposed to build it by subscription, and calling a meeting for 20 December at which proposals would be received " . . . for the delivery of coal and lime at Talybont; of filling a subscription for the intended RAILWAY, and other necessary arrangements".

Penry Williams, who was to be associated with the Brinore Tramroad until 1874, took the chair at the meeting at which it was announced that Jonathan Dixon and George Overton had rented a colliery at Brynoer from Benjamin Hall (this was not strictly true as the lease was not signed until July 1814), and were prepared to deliver coal and lime at Talybont at much reduced prices, and were further prepared to lease the proposed tramroad for five years at seven per cent of the cost of construction provided that it did not exceed £14,000. It was announced that Hall was prepared to guarantee the payment of the rent by Dixon & Overton, subject to the subscription being filled by 21 February 1814. A resolution was passed that " . . . as such a Railway and communication will be essential to the Interests of the Counties of Brecon, Hereford and Radnor a subscription be opened for carrying the same into effect". The meeting was adjourned to 21 February, and Church was authorised to receive subscriptions in the meanwhile.

Hall was unable to attend the meeting in February owing to indisposition, but it was reported that £11,800 had been subscribed. The shortfall, of course, raised a doubt as to Hall's willingness to fulfil his guarantee, and Church was instructed to write to him in the following terms:

> "On the supposition that Mr. Hall will continue his guarantee for the payment of 7 per cent and for securing his Road to the Public as a Public Road from the 8 mile post to its termination with Liberty of being joined by any Roads that may be required to be connected with it (Mr. Hall receiving the tonnage on his own Road) . . . and that Mr. Church inquire of Mr. Hall whether he will complete the subscription required".

Thus it was revealed that Hall intended to benefit directly from the tramroad by a continuation to Rhymney over land already under his control, and over which he did not require the permission of the B&A to construct a tramroad.

The next meeting was held on 14 March when it was announced that £13,000 had been subscribed, and that a further £1,000 had been promised should it be required. The meeting proceeded with the normal business of an inaugural meeting, appointing Peirce, Williams & Co. of Merthyr Tydfil, bankers, to act as treasurers, selecting committees, and laying down procedures for meetings. It was decided that the title of the company should be "THE BRINORE TRAM ROAD", which, whilst it may offend lovers of Welsh orthography, was nonetheless the correct style of the company.[12] John Llewellin was appointed to make the necessary payments to the contractors, and to oversee the construction generally.

The subscription list included Benjamin Hall (£2,000) who would gain access to the canal for his Rhymney ironworks; Jonathan Dixon, Overton's partner in the Brynoer Colliery (£3,000), with his obvious interest in providing a route by which to send coal to Brecon: Sir William Keppell (£500), who evidently hoped that his interest in the Trevil Rail Road might benefit; Rev. Archdeacon Davies (£500), who was a shareholder in the Hay Railway, and foresaw traffic for that concern; Fothergill & Co., of the Sirhowy Ironworks (£500), who doubtless saw a possible further outlet for their goods; Moses Moses and Rosser Thomas (£100 each), who were already mining coal near Brynoer Patch; John Llewellin (£100); Peirce Williams & Co., Merthyr (£500), the company's bankers; and Samuel Church (£200), the company's solicitor. The remaining £5,500 was from local people who could see the advantage of breaking the monopoly to which the district had been subjected for the past decade; but it is hard to see what Matthew Wayne, of Gadlys Furnace, Aberdare, stood to gain other than a small investment. A notable absentee from the list was George Overton, who had lost a great deal of his capital in the failure of the Hirwaun Ironworks in 1812.

A first call of ten per cent had been made at the inaugural meeting, and on 12 April a further ten per cent was called. The latter meeting examined and approved Overton's offer to contract for the construction of the line in the sum of £12,000, with a completion date of 1 May 1815, and Church was instructed to prepare the contract and the guarantee to be signed by Hall, in order to have everything ready by 16 May. There was considerable delay before either of the documents could be executed, and Church reported to the meetings of 24 May and 6 August that he had not, even then, received the approval of the deeds by the solicitors acting for the other parties. At the latter meeting, in order to get matters moving, Church proposed that he should join with Dixon & Overton in a bond for the due performance of Overton's contract. The meeting approved this arrangement, and the executed deed was passed to Penry Williams to hold on behalf of the shareholders. It appears, however, that work had already started, as it was decided to make another call of thirty per cent, "... owing to the very advanced state of the Road ...", but a proviso that no further call should be made unless Llewellin was satisfied that progress justified it was added.

Meanwhile, Dixon was making preparations for the traffic expected at Talybont. On 27 April 1814, he was granted permission by the B&A to construct a basin for turning boats, and a month later he received permission to "teem" the earth thus removed on the tow-path side of the canal. He also erected limekilns and sheds, a granary (he farmed Wenallt Farm, Talybont), two dwelling houses and a counting-house.[13]

In July, Dixon & Overton leased the coal under sixty acres of Brynoer Patch, near Rhymney, from Hall, the lease to become effective from 1 January 1815.[14] They were granted a term of thirty-one years, at a sleeping rent of £400 a year, with a galeage of 1s. per ton on eighty out of every one hundred tons raised, and 4d. per ton on the remaining twenty tons; thus for each one hundred tons raised they were to pay a galeage of £4 6s. 8d., and unless 9,230 tons were raised in the year, the minimum rental of £400 became payable. They were also obliged to produce a minimum of twenty-five tons daily. In return for these rather high charges for the coal, Hall had already undertaken to provide a tramroad, on which the quite moderate tonnages of 1½d. per ton per mile on the eighty tons, and 1d. per ton per mile on the twenty tons were to be charged. Hall undertook not to grant better galeage or tonnage to any other party, and to purchase any "mine" (iron ore) that was found with the coal at 12s. per "dozen" of thirty-six hundredweights, a somewhat archaic measure derived from the carrying capacity of twelve horses with panniers.

By 2 September 1814, "the very advanced state of the Road" rendered a further call of twenty per cent necessary, and in mid-October "the near completion of the Road" justified another call of twenty per cent.

The remaining ten per cent was called on 25 April 1815, when Llewellin reported that:

> "... the Road contracted for by Messrs. Dixon & Overton is now nearly completed—the Main Road quite finished—That several partings are already laid & the Plates required for the whole delivered upon the Work—and also every Thing necessary to be done will certainly be completed according to their Contract within three weeks from this date—and that at present it may be worked upon without detriment—the additional number of partings being only required for an increased Trade ..."

The fences were nearly finished and the expenditure of only £100 would be necessary to finish everything. The report continues:

> "That the remaining part of the Road to be made by Mr. Hall is all laid within about Two Hundred Yards & will be ready to be worked upon by the First of May next ..."

Llewellin also considered:

> "... the whole of the Work had been done in the best possible manner and that it is in his opinion one of the most substantial Roads in the Kingdom."

As the treasurers, Peirce Williams & Co., still had £984 in hand, it was ordered that the committee should meet on 23 May to settle all outstanding accounts, and Llewellin was given the job of finding a suitable place for the erection of a toll-house, and instructed to produce a plan of it at the next meeting. The meeting concluded with the passing of the following resolution:

> "Messrs. Dixon & Overton having proposed to deliver in Boats at the public wharf at Brecon, Sixty or more tons of coal per diem at Twelve Shillings per Ton and any quantity of Lime, not less than a Boat Load per diem likewise in Boats at the same place for Fourteen Pence per Barrel—Resolved that the Cordial Thanks of this Meeting be given them for thus enabling the Proprietors of the Brinore Tram Road to redeem every Pledge that they have originally or at any subsequent period held out to the Public".

Doubtless the tramroad was opened, as Overton had promised, on 1 May 1815, or shortly after, but neither the press nor the minute book recorded the fact at the time.

II

Little time was wasted in paying for the land used for the tramroad, and the conveyances were completed on 2 May 1815, all passing ownership to Samuel Church, as trustee for the shareholders.[15] The lease of the tramroad to Dixon & Overton was executed on 15 May, Church acting in the same capacity.[16] Under this, they took the line at an annual rental of £896 for five years, and undertook full maintenance. They were also to keep the tramroad public, and open to all persons on payment of the authorised toll.

The maintenance clauses of the lease were explicit, and throw light on the construction methods. As new plates were required they were to be laid on "iron dove-tailed sleepers with flat stones under the end similar to the Tram Plates now laid down". The lessees were to keep the tramroad, " . . . coursed Nine Feet wide and Six Inches thick with Broken Stones and keep the space between the Plates filled with Small Stones or Gravel . . . " The boundaries of the tramroad were to be delineated by "a good quickset [hawthorn] hedge except at the head of Dyffryn Crawnon, where they were to maintain a sound wall, and where neither method was practicable, fences were to be erected". With regard to trade, the lessees undertook to supply not less than forty tons nor more than one hundred tons of coal daily at Talybont, there to be sold at not more than 11s. per ton; similarly, they were to supply 300 barrels of lime (about thirty-eight tons) daily to be sold at 1s. per barrel.

Llewellin was given the contract for the construction of *two* toll-houses at the meeting of 13 June 1815, to be built to his specification. Unfortunately, the minute book does not quote their locations, but one has been located at Talybont at SO.1138 2241; the other has not been located but was probably in the vicinity of the eight mile post in Trevil.

Early in 1816 it was mooted whether it would be advantageous for the company to obtain an Act of Incorporation from Parliament, but it was decided that as the tramroad had been constructed under the B&A Act, the same obligations and privileges must apply equally to them as to the canal company.

Accounts for the construction of the line were completed by the meeting of 4 March 1816, although it must have been known by the previous May, when Dixon & Overton's rent was fixed, that the total cost had been £12,800. To break down this sum is a little difficult, as it is not known if the conveyances found (note 15) represent the total of the land the company had found it necessary to purchase. Using the figures quoted in the minute book under this date, and assuming the deeds mentioned to represent the total expenditure on land, the accounts must have read, roughly, as follows:

Cost of actual construction	£12,126.11s. 4d.
Legal expenses	£396.11s. 7d.
Purchase of land	£241.11s. 2d.
	£12,764.14s. 1d.
Retained for contingencies	£35. 5s.11d.
	£12,800. 0s. 0d.

With the tramroad completed, and let on terms guaranteeing seven per cent on the sum expended, and £200 of the subscription still in hand, prospects must have looked rosy, particularly as under the lease to Dixon & Overton no part of the maintenance costs could fall on the shareholders. Nonetheless, it was considered necessary to remind the lessees that the company would expect the rent to be paid on the first Monday in May.

The meeting in May had to be adjourned, a quorum not being present, and the June meeting revealed a somewhat unusual state of affairs. Instead of Dixon & Overton paying the rent for the tramroad in the normal manner to the treasurers, it was expected that they should pay it direct to the shareholders in proportion to their holdings, as the dividend. In this they had defaulted, stating that whilst they were prepared to pay those with under £200 invested immediately, they would be unable to pay the remaining shareholders until 2 February 1817, proposing to issue promissory notes to be then payable, with interest to the date of payment. The meeting had, perforce to agree, the only alternative being to call on Hall to pay under his guarantee.

It is difficult to visualise this method of dividend payment being satisfactory. Dixon & Overton were charged a rental of £896, representing seven per cent on the cost of £12,800. The amount subscribed was £13,000, and to pay seven per cent on the full investment would have required a rental of £910 per annum.

In the following year the same difficulty arose, and the meeting of 18 December 1817 noted that the dividend payable in the previous June was still unpaid. The solicitor was instructed to demand immediate payment.

A futher complication arising from this unusual method of paying dividend was revealed at the meeting held on 10 December 1819. In common with many other concerns, the Brinore Tram Road Co. discovered that it was one thing to obtain subscriptions undertaking to supply capital, but it was quite another matter to ensure that obligations were met. The treasurers were instructed to call forthwith upon the defaulters for the immediate payment of the arrears,

> "... and that they also call upon the defaulters who have received Seven Pounds per cent upon their full Subscriptions to refund the Interest so overpaid them ..."

No indication is given in the minute book as to the means by which Dixon & Overton were supposed to know who had not paid in full.

The extent to which coal from the Brynoer Colliery affected the B&A, and, of course, the Brecknock Boat Co., can be gathered from the canal company's statement of accounts for the year ended 26 April 1816. At the foot of the printed copies sent to shareholders the following note appears over the signature of B. A. Griffiths:

> "I beg leave to observe that in consequence of the depressed state of the Country, and the loss of Tonnage by the short freight from Talybont, the amount of Tonnage this Year is short of that of last Year the Sum of £392 2s. 6d. ... "

Assuming that this was caused entirely by Brynoer coal taking the place of that from Clydach, this would represent a loss of trade to the BBCo. of some 4,000 tons. How this came about, and incidentally accounting for Dixon & Overton's inability to pay their rent, is summed up in a letter from Thomas Ellwood, mineral agent to the Boat Co., to his principals dated 3 June 1815:

> "As to the Brin Oar Coal should the country be inclined to favour them I should was it my case imidiatly reduce the Colliery and by no means make any further Reduction in Price if they chuse to sell Coal at less than Prime Cost why should you do the same . . . "[17]

Dixon & Overton were not the first to land themselves in difficulty by attempting to break into an established market by undercutting the existing supplier!

The partnership between Dixon and Overton had been formalised by a deed of 19 November 1816, and had been intended to endure for thirty years, from 1 May 1815. Late in 1819 Dixon decided to retire due to failing health. He had brought into the partnership a considerable amount of property which he held jointly with his wife, as tenants in common, including the wharf at Maesmawr, Talybont, the adjacent kilns and the counting-house, all of which were leased to the partnership at a rent of £150 per annum. Overton took as his new partner Edward Gervase Scroop on 27 January 1820, the new partnership to continue to lease the Brynoer Colliery and the property brought into the concern by Dixon at the same rent as before. Until his capital had been repaid, Dixon was to receive a royalty of 9d. per ton on best coal, and 6d. per ton on "inferior" coal together with 3d. per ton on limestone, all amounts to be calculated on the quantities carried on the tramroad; as he was a farmer, he was to be permitted to take away "the waste lime and clearings up" on the wharf. The agreement between Dixon and Overton & Scroop appears to have been drawn very loosely,[18] and after Dixon's death on 23 July 1820,[19] his widow indulged in a long drawn-out law-suit with Overton & Scroop over various matters associated with it. This was still continued after 1843 between Hoper Dixon, Jonathan's son, and Overton's son, George, after the principals were dead.[20]

The original lease of the tramroad was due to expire on 1 May 1820, and on 2 February, Llewellin was deputed by the company to survey the line to ascertain that the maintenance clauses had been complied with. At the same meeting, Overton & Scroop offered to take a further lease but their proposed terms were unacceptable, and the meeting decided to advertise in the *Cambrian* that the letting of the tramroad would be decided at a meeting to be held on 28 April, in the hope that a better offer might be forthcoming.

Llewellin's report was not available on 28 April, but an unnamed person reported that "Mr. Overton is busily employed in repairing the Tram Road which he expects will be completed by the First of June next". Feeling that they could not rely on Overton, a committee of three was appointed, to be attended by "some competent person", to examine the line as soon as possible after the date set by Overton. Despite this, Overton & Scroop were granted a new lease for one year, which could be renewed for a further two years, but could be determined on six months notice. This was on the same terms as the previous lease with regard to maintenance and deliveries, but the rent was reduced to £768 (six per cent of £12,800). A reliable surety was also required for the payment of the rent.

Other business at the meeting included a report that there was a sum of £200 in the hands of the treasurers surplus to immediate requirements, and it was resolved that this should be distributed amongst the paid-up shareholders in proportion to their holdings. There was also a request from newcomers to the scene, Messrs. Powell & Williams, who were to provide additional competition in the coal trade, for the provision of a weighing-machine for their use. The small committee who had examined the line was instructed to find a suitable place for the machine, which was to be

Brinore Tramroad Office, Talybont on Usk.

installed at the expense of Powell & Williams. This company, trading as Powell & Co. was later to prove a very sharp thorn in the side of the tramroad company.

The competent person chosen by the committee of three was none other than Richard Oliver, Agent and Clerk (and general factotum) to the Hay Railway, which had been opened in 1816 from Brecon to Hay and Eardisley. A preliminary report was given to the committee on 5 September stating that "the Road, Partings and the Fences are not in repair" and Church was given the job of writing to Overton on the matter.

Oliver's full report was presented to the company at the meeting of 24 October, and was critical both of the standard of the original construction and the maintenance. He considered that whilst the road was in a tolerable state of repair in most places, the plates were badly sunken at Penrhiwcalch, and generally the plates were too light for use on a public road, being only two-thirds of the weight normally employed; he approved of the use of iron sleepers as being "of great service in keeping the Road to the proper gauge and in good order", but thought that larger stones should have been placed under the ends of them as he had noticed in places that one side of the track was lower than the other due to the stones sinking. He was especially critical of the paucity of passing places, pointing out that it was normal to provide three or four in every mile whereas there were only thirteen in the whole eight miles of the Brinore. The fences he thought had too few "quicks" planted, and he had found difficulty in places to distinguish between the company's property and the surrounding country. At the head of the Dyffryn Crawnon valley he considered the situation to be very dangerous. At this part, the tramroad was constructed on a shelf cut into an almost vertical hillside, and it is easy to imagine his alarm, used as he was to the Hay Railway carried entirely through low-lying agricultural country. His concern on this point was possibly justified, and indeed, the company received a claim in respect of two horses which fell off the edge in 1855. His report concluded that:

> "There appears to be a tolerable trade on the Road at present, and when we look at the situation of this, with other Roads in the Neighbourhood, it is naturally to be supposed it will be very much increased within a few years—As such I recommend that the permanency of the Road should be particularly attended to."

On reading this report, the meeting instructed their solicitor to write to the former lessees for the fulfilment of the terms of their agreement. With Dixon dead, this, of course, meant Overton, who as contractor for the original construction of the line would have been wiser had he made it in the most substantial manner possible. Church was also instructed to write to Overton & Scroop calling upon them immediately to provide a surety for the completion of their contract.

Overton & Scroop continued to ignore the company's demands for a surety, and in May 1821, it was decided to inform them that if such was not provided within the next fortnight, the lease would be considered as terminated, and they would be considered liable for the rent up to the date of cancellation. They were, however, let off the hook by John Llewellin, who on 7 June pursuaded the company to continue the lease without security.

The proprietors seem to have been impressed with Oliver's report and at the May meeting passed a resolution that the committee should consider themselves empowered to employ "a competent Person occasionally to overlook and examine the state of the Road and report the same."

William Powell & Co. appear to have commenced trading over the Brinore in the first half of 1820, and in April they applied to the B&A for the weights as shown by the Trevil Machine to be the basis of the tonnages charged to them on the canal, and their proposal was accepted.[21] Early in 1821, they applied to the canal company, and were granted permission to construct a wharf at Talybont, adjacent to the waste weir (SO.1144 2251), and to construct a line of tramroad to connect with the Brinore; they were also permitted to lay a tramroad from their wharf to the existing Maesmawr Wharf (made by Dixon) on the "off" side of the canal, and their offer to pay a quarter of a mile tonnage on all goods carried along this stretch was accepted.[22] It is probable that this line constructed by Powell was the only branch ever made from the Brinore.

Two months later, Powell & Co. had bigger ideas, when they applied to the assembly of the B&A for a tramroad to Talybont to serve their limestone quarries in the parish of Llangynidr. The proposal was turned down, but they were informed that the canal company would put no obstacle in the way of their making it for themselves.[23] The location of Powell's quarry is not known, and it is considered very unlikely that they carried on with the plan.

Probably Powell & Co. completed their construction work at Talybont by August 1821, as on the 28th of that month, a fresh lease with Overton & Scroop was concluded under which their rent was reduced to £640 (five per cent of £12,800), and they were relieved of the obligation to supply coal and lime at Talybont. It seems that Powell & Co. took over this trade leaving Overton & Scroop free to concentrate on the trade in Brecon and at their depots on the Hay Railway. The new lease, which was to last for two years, retained the maintenance clauses of the previous lease, and was to be renewed at the end of the period on the lessees giving six month's notice of their wish to continue. In order to reduce the wear on the track, it was ordained that the maximum load to be carried in any tram should be thirty hundredweight, which was to the advantage of Overton & Scroop.

In November 1822, there was a further variation in the lease on terms proposed by Overton &

Scroop, whereby they were to hold the tramroad for seven years at the same rental payable half-yearly, but they could, on giving six month's notice, terminate the lease at the end of any yearly period. The conditions with regard to maintenance were unchanged, but they were expected to put up the tramroad into good order immediately, and to keep and leave it in the same condition. The Brinore Company, however, reserved to themselves the right to take over the tramroad on giving six month's notice if "... the Profit on the said Road shall be equal to the Sum necessary to relay the same where necessary ..." With an eye to the future, it was ordained that any renewal of plates should be made with heavier ones.

There was a surprise on 27 May 1823, when Church announced to the meeting that Overton & Scroop had already given notice to terminate the lease on 6 June, unless the rent was reduced to £512 (four per cent of the cost of construction). The shareholders would not entertain such a reduction, and a resolution was passed that they would not let the tramroad to Overton & Scroop at less than five per cent, and that the line should be inspected to ascertain that it had been properly maintained. It was also decided that should there be no other takers for the line the shareholders should take it into their own hands and "... that a proper person be employed by them to superintend the same and receive the Tolls".

The shareholders met on 11 July and received the report on the state of the tramroad by Theophilus Rhosser Jones, described in the minutes as "Tram Road Agent". Details of the report are not quoted, but Church was ordered to apply to Overton & Scroop to get the repairs effected, and to offer them a further week in which to decide to carry on running the tramroad at five per cent. Should they decline, a committee appointed at the meeting was to take over the running, and employ a superintendent to look after day to day matters. No action had been taken to put the line in order when the next meeting took place in September, and Church was instructed again to call on Overton & Scroop to effect repairs, failing which they were to pay £159, the amount Jones considered necessary to put the line in order.

A week later, on 13 September, it was decided to sue Overton & Scroop should they not pay up within a fortnight, and the taking over of the management of the line by the company was advanced by the appointment of Anthony Powell "... to act as Clerk to this Company at a salary of one Guinea per week (which sum is to cover all Expences and Extras) & he is to dedicate the whole of his time to This Concern." His duties were described as to receive and give an account of all tolls and to pay the amount of his receipts fortnightly to the committee, and to superintend the line generally.

No payment had been received from Overton & Scroop by 14 August 1824, and Church reported that he had commenced an action against them. On the suggestion of Philip Jones, the Abergavenny banker, who had acquired an interest in the share originally held by Fothergill & Co.,[24] it was decided that the action should be called off if Overton & Scroop would, within a fortnight, pay £50. The suggestion was rejected by Overton, who was present, but no reason for his refusal is recorded. Possibly the partners had a counter claim against the company, as on 10 November 1825 Church was ordered to sue Overton & Scroop for an outstanding balance of £15 17s. 2d. only; possibly Overton had a change of mind, and the smaller amount was something entirely different. The smaller amount was still in dispute a year later, and Church and Hoper Dixon (son of the late Jonathan) were instructed to look into the matter fully.

November 1826 saw a small change in the administration of the tramroad when it was decided to hold the meeting annually in October at which the accounts and a report on the state of the line should be presented "... in Order that Directions may be given respecting the Repairs ... it being the wish of the Proprietors that the Road shall be kept in as good a state of Repair as possible". It was also ordered that all trams using the line should have the names of the owners and a number painted on the side.

The first dividend to be declared after the assumption of management by the company was authorised in October 1827—a modest four per cent. However, it is probable that some money had been spent on repairs and renewals, as Hoper Dixon reported that, taking it generally, the tramroad was in a better condition than at any time for the last eight or ten years; he did admit that he was afraid that some expense would be incurred in the next two years on the section from Penrhiwcalch to Talybont, which was the steepest part of the line, as the placing of shoes under the tram wheels to act as brakes caused greater wear on this part. He concluded his report:

> "I consider Messrs. Overton & Scroope's Clk [Clerk] blameable in allowing a considerable quantity of soil to be lodged from the lime Rocks in the Tram Road particularly after Messrs O & S had been repeatedly cautioned not to do so".

Poor 'O. & S.'—their copy-book was so blotted that they could always be criticised!

A year later, in October 1828, the dividend had risen to four and a half per cent, the highest ever paid; in fact, whilst there must be a number of meetings omitted from the minute book, it appears that this figure was only paid in 1828, 1829 and 1831, as far as the record goes.

At the October meeting a proposal was received from Summers Harford of the Ebbw Vale and Sirhowy Ironworks that a connecting tramroad should be constructed between the Sirhowy Works and the Brinore. The Brinore was unable to make up its mind, and appointed Hoper Dixon to investigate the possibility of any profit arising to the tramroad company from this source. His report has gone unrecorded.

Harford, who has acquired one share in the Brinore in August 1831,[25] returned to the attack in May 1832, but refused to make a written application, stating that he had already sent a communication to Church. The meeting was adjourned until Church should re-convene it should he find the letter, but no further action is recorded as far as the Brinore was concerned, and Harford seems to have turned his attack onto the B&A and attempted to induce them to build a new tramroad for him. The Brinore wrote to the B&A protesting against this line in October 1832, and after much delay the B&A refused to make it in 1839, by which time the impending bankruptcy of Harford, Davies & Co. was common knowledge.[26]

Peirce, Williams & Co., of Merthyr Tydfil were declared bankrupt in 1824, and although they eventually paid 18s. in the pound[27] the Brinore had to find other bankers, and from 1828 the name of J. P. Snead appears in the list of shareholders attending meetings. He was a partner in the firm of Bromage & Snead, of Monmouth, who had a branch bank in Brecon[28] where, from 1831 dividends were declared payable. The firm had been in low water in 1825 but seems to have regained liquidity very rapidly.[29]

With a view to increasing the traffic on the tramroad, a reduction of ½d. per ton per mile was made in the tonnage on coal in February 1832, and a charge of 1½d. per ton per mile was made for several years until the matter was brought up at the assembly of the B&A in October 1838. The Clerk of the B&A, now James Peirce, who had succeeded Griffiths in 1827, was instructed to write to the Brinore and point out that under the B&A Act tonnages were to be equal on all parts of the system, and that such a reduction was illegal. In January 1840, the B&A received notification that the tonnage had been adjusted to the same figure (2d.) as charged on the canal. The Brinore had, indeed, discussed the matter at their meeting in January, but George Overton, junior (Overton had died in 1827) raised the point that it might be considered that freighters were liable for the extra tonnage throughout the period in which the tramroad had charged the lesser amount. He took counsel's opinion on the matter, which it must be assumed was in the freighter's favour as there is no record of any liability for arrears. The motion to raise the tonnages was passed in October 1840. A legitimate way of effecting a slight reduction was accepted in May 1841, on the suggestion of Hoper Dixon, when a drawback or reduction was made in the tonnages charged on small coal for lime burning of ½d. per ton per mile on the first 2,000 tons of such coal carried. This small coal commanded a lower price, and due to the seasonal nature of the lime trade, payment for it was frequently outstanding for six months or more, and consequently, payment of tonnage accounts delayed. No doubt it was hoped that having to pay a lesser amount might make payment a little more prompt.

Difficulty was being experienced with the clerk, Anthony Powell, who, in October 1831, failed to produce vouchers to support his accounts. A year later, he failed to produce the accounts in time for the annual meeting, and a resolution was passed that it was the opinion of the meeting that he was overpaid at 25s. a week (he must have had an unrecorded rise at some time), and his salary was reduced to 18s. a week as from the following January. The resolution concluded that he should be informed that " . . . unless he conducts himself more soberly, he will be discharged from his situation as Clerk". There are no more recorded complaints of his behaviour, but he must have fallen from grace at a later date, as the minutes for 18 January 1840 ratify the appointment of Richard Ball as clerk from 1 January 1839, at the same salary of 21s. weekly as originally received by Powell.

The condition of the track had evidently been considerably improved by December 1835, when the maximum permitted weight for trams and their contents was increased from forty to forty-five hundredweight.

III

Two advertisements appeared in the *Hereford Journal* of 8 October 1845, each calling for a general meeting on 27 October; the first was to elect a new solicitor and trustee in the place of Samuel Church, deceased; and the second " . . . to consider the propriety of disposing of the said tramroad . . . " The meeting was duly held, and the former matter resolved into a contest between George Overton, who was a barrister, and David Thomas, a Brecon solicitor, in which Thomas was elected by a majority of ten votes. On the second point, the meeting was adjourned for a month, to enable the shareholders to judge offers which had been received from the Hereford & Merthyr Tydfil and the Welsh Midland Railways.

The meeting of 24 November confirmed that it wished to dispose of the tramroad, and decided to accept the offer of the H&MT to pay £105 for each £100 Brinore share, to be paid within two months of the company obtaining its Act, and a deposit of ten per cent of the amount to be paid was to be lodged with Bromage & Snead, the tramroad bankers, within two months of the signing of the agreement. On 8 December, Thomas was instructed to prepare a contract to be forwarded to the H&MT solicitors for their approval. Later in the month, Thomas reported that he had been informed, verbally, by the local solicitor to the railway company, that the H&MT would be unable to fulfil their engagement, and he was instructed to write to the London solicitor acting for them to ascertain the position, and should his reply be indefinite, to call off the deal. The reply was reported to the tramroad company on 12 January 1846. It consisted of a copy of a resolution by the H&MT directors to the effect that they did not consider that they were in a position to conclude the agreement with the Brinore, and that they trusted that they would be in a position to re-open negotiations in the following Session of Parliament. The shareholders decided that there was no point in pursuing the matter any further, and thus faded the first of many attempts by the tramroad company to dispose of its property.

Brinore Tramroad: Plate rails and tie-bars.

The meeting also received a report from Ball that he had received notices of intent to apply for Acts from the Brecon & Merthyr Tydfil Junction Railway, and a further notice from the Welsh Midland. He was immediately authorised to issue notices of dissent to both proposals on behalf of the tramroad.

Realising that the value of the tramroad to another concern could be enhanced, it was resolved on 19 October 1850 to accept the offer of one of the proprietors, T. P. Price,[30] to allow his surveyor, a Mr. Habakkuk, to examine the line with a view to altering the curves and relaying with wrought iron plates, to make it suitable for locomotive working. His report was received at the November meeting, and whilst he was thanked for his "very clear, able, and practical report", it was decided that in the present state of its finances, the company could not afford to pay the £6,200 estimated to be required, and the idea was allowed to rest.

The company's funds were very low by September 1854, when it was reported to the meeting that owing to losses by bad debts (for which no provision had been made) and the repayment of their overdraft to Bromage & Snead, only sufficient money remained in hand to enable a dividend of one per cent to be declared. Tighter financial control was ordered and the Clerk instructed to produce his accounts quarterly for examination by Hoper Dixon.

By the following June it was reported that the Talybont Coal Co., who had taken over the business of William Powell & Co., owed £500 in respect of past tonnages for which dishonoured bills of exchange were held, and a further £296 9s. 9d. was owing in respect of tonnages for the quarter ended on 31 December 1854. Thomas was instructed to write them, and if the reply was unsatisfactory, to take the matter to court. Generally, the company's accounts were in a bad state, and it was decided to call in T. Chandler Perks, of the Hay Railway, to inspect the accounts; to examine the state of the tramroad, and to report on them quarterly. His remuneration was not to exceed £10, but in view of the continuing nature of his mandate, it must be assumed that this was to be an annual payment.

Another attempt to sell the tramroad was initiated in June 1855, when on hearing that the Rumney Railway (the 'Old Rumney' from Rhymney to Pie Corner, Bassaleg, near Newport) proposed to turn itself into a passenger carrying railway, it was resolved that the Chairman, Rev. T. J. Powell, should approach Crawshay Bailey (who held one-sixth of the capital in the Rumney), with a view to finding out if it was their intention to extend their line into Breconshire. If not, he was to contact the Rhymney Railway, which had obtained its Act in the previous year, with a similar question.

Perks reported to the meeting of 4 July 1855 in a manner which was hardly favourable to the Brinore. He considered that 4,598 yards of track was either in need of relaying or re-ballasting. The line at the head of the Dyffryn Crawnon he pronounced as " . . . in a bad state which would be pronounced on the Hay Railway as quite impassable . . . ", and he estimated that it would cost £405 14s. 4d. to put the whole road in good order. He suggested that a saving of between 7d. and 9d. per yard could be made when relaying by the substitution of larch sleepers for the cast iron ties, and if wood was used, such good ties as were taken up could be used to repair other parts. Of the three men employed to deal with maintenance he had found only one to be working, and he was unsupervised and entirely without instructions; to his enquiry as to the whereabouts of the other men he was informed that they were trimming hedges, which Perks considered " . . . not to be the work that is necessary or required at this time of year . . . " On the accounting side, he had met with almost complete obstruction, Ball having refused to give up the books and papers requisite for a proper investigation. He reported an utter lack of system:

> " . . . and the books so called Day Book and Ledger are not such as a public company ought to expect—the Ledger (so named from its external appearance) is merely a summary or recapitulation of the Day Book . . . it does not carry forward any balance, it does not show one receipt or payment, neither does it exhibit the financial position of the Company".

Ball was given until the ensuing 8 August to get as much as possible done to the track, and ordered to hand over all financial records to enable Perks to prepare a statement of the company's affairs.

When it met on 8 August, the company was given a summary of the accounts for the last five and a half years prepared by Perks, and he was requested to hand a copy of the cash account to Ball to enable that gentleman to explain the discrepancies. It appeared that Ball had made no attempt to put the line in repair, and he was ordered to attend the meeting to be held on 23 August to explain his behaviour. He did not put in an appearance at the meeting, nor was a report on what he had done towards repairing the tramroad forthcoming, and he was dismissed instantly. As they now had no clerk, Perks was asked to fill the position until it was possible to find a suitable candidate, and to assume full charge of the tramroad temporarily.

In the meantime, T. J. Powell had heard from Crawshay Bailey to the effect that the 'Old Rumney' had no intention of extending their line into Breconshire, and he had carried out the other part of his mandate and contacted the Rhymney Railway. This had elicited a suggestion that deputations from each company should meet at any early date for an exchange of views. The meeting of 8 August appointed a deputation consisting of David Thomas, the solicitor, T. J. Powell, Hoper Dixon and William Llewellin, son of John, who had succeeded his father as

Agent to the Hall family, to represent the Brinore, and authorised them to pass on any information in respect of the tramroad.

The deputation reported back to the meeting of 23 August 1855 that the Rhynmey Railway required a plan of the tramroad, and Llewellin was given the job of tracing the one prepared by Habakkuk in 1850, or should that have been lost, of employing a surveyor to prepare a new one. The deputations met for the second time on 12 September, but the RR party refused to transmit the proposals to their directors unless a further plan showing the gradients from Penrhiwcalch, down the Collwng Valley to Talybont was produced; evidently it had been hoped to conceal that this part was steeply graded, and it was only after adjourning their meeting from 19 September to the 25th that they made up their minds to comply with the RR request, and instructed the surveyor, Isaac Davies, to amend the plan to show the radii of all curves and the gradients. Davies forwarded the plan as amended to William Gascoigne Roy, the RR solicitor, and there matters seem to have rested. On 22 May 1856, it was brought up that no reply had been received from Roy, and Thomas was ordered to write to him. The reply was received on 24 May from Roy's clerk to the effect that Roy was too ill to attend to business; a further reply of 10 June stated that he was still indisposed, and the meeting of 12 June was adjourned until Thomas should re-convene it on receipt of the final answer.

No further discussions are recorded in the minute book, in fact, the date of the next meeting that is recorded is 7 September 1857. There are, however, papers in Gwent Record Office[31] headed "Remarks on the advantages of a Railway from Rhymney to Brecon", which are, unfortunately, undated but which undoubtedly refer to this scheme. Whilst they put a very rosy hue on the possible returns from such a line, they reveal that it would be at such a cost as to frighten off the Rhymney whose small capital of £100,000[32] would have been very nearly swallowed up in the estimated cost of building the twelve miles proposed, using five miles of the Brinore, for £89,390. Another paper, dated 12 September, written on the back of a Cardiff Arms Hotel letter heading gives the gradients of the Brinore as:

> "From Talybont
> Gradient 1 in 25. 40 ch.
> 1 in 21. 80 ch.
> 1 in 26. 80 ch.
> 1 in 22. 80 ch.
> 1 in 76. 240 ch. Ascending from Talybont
> 1 in 55. 120 ch. Descending"[33]

which, probably, also awed the RR deputation. Thus died another effort for the sale of the tramroad.

IV

It will be remembered that the meeting of 23 August 1855 had appointed Perks to be temporary Clerk to the tramroad. A week later he was appointed to the position permanently at a salary of 25s. weekly with the gratuitous use of the cottage, office and garden formerly occupied by Ball. He was required to provide a surety for £200 and " . . . at all times during his absence to keep and supply an efficient person on the spot for the due performance of the duties of Clerk . . . ", from which it can be inferred that he was to retain his position with the Hay Railway; this is confirmed by the fact that a T. C. Perks described himself as 'Manager of the Hay Railway' when that concern was taken over by the Hereford, Hay & Brecon Railway in 1860.[34] Perks also undertook to supply all necessary horse labour for the tramroad for the sum of £10. At this price the 'horse labour' can only have been for use during repair work, as the normal charge for horse haulage was 2s. 6d. per day.

By April 1856, Perks had so proved his worth to the company that he was given *carte blanche* "... to pay all necessary expenses and disbursements for the repair and general management of the Tram Road". As his accounts in respect of horse labour showed that he had made a loss in providing this service, he was voted an extra £5 annually from the start of the contract. His appointment was undoubtedly a wise move, and it was unfortunate from the tramroad company's point of view that he had not been available before the decline of its fortunes set in.

Almost from its inception, the Brinore had experienced competition from the Old Rumney Railway (a plateway) which had, from about 1828 provided an outlet to Newport for the products of the Union Iron Co. and its successors the Bute Iron Co. (1825) and the Rhymney Iron Co. (1835). The Monmouthshire Railway & Canal Co. had forbidden the use of animal haulage on its tramroads from 1 August 1849, and as the Rumney used their track from Pie Corner to Newport Docks, locomotives had become the order of the day on that line also. When, in 1860, the whole of the MRCC tramroads had been converted to railways, the Rumney had, perforce, to keep in step. The Rhymney Railway giving access to Cardiff Docks over Taff Vale Railway metals was opened in 1858 and provided a further outlet for coal and iron from the Rhymney area. How the Brinore was affected is shown in the following traffic returns. These record only the gross tonnages received and take no account of any outgoings.

	1850	1857	1858	1859	1861*	1862*
Coal	752. 1. 0.	202.15. 7.	176.16. 0.	83.16. 2.	326.19. 2.	255. 1. 9.
Lime and Limestone	184. 2. 2.	124. 2.11.	95.10. 9.	85.15. 7.	84. 3. 8.	71. 5. 1.
Pitwood		196. 5. 6.	180.10. 9.	117.14.10.	133.19. 7.	118.15. 2.
All other goods	243.18. 1.†	55.16. 4.	43. 6. 2.	43. 3. 3.	22. 1.11.	24. 7.11.
	£1,180. 1. 3.	579. 0. 4.	496. 3. 8.	330. 9.10.	567. 4. 4.	469. 9.11.

* The 1861-62 figures include traffic for the construction of the Brecon & Merthyr Railway.

† The 1850 figures taken from GRO. ref. Man/G/1-006 do not make any distinction in the classification "all other goods".

It has often been stated that it was the Brecon & Merthyr Railway that killed the Brinore, but the foregoing figures, which are the only traffic returns to be traced, show that the railway can only have given the *coup de grâce*.

We left the Talybont Coal Co. owing the Brinore £796 9s. 9d. in June 1855. The following month Thomas reported that he had only received £200 from them, and he was instructed to take proceedings to recover the outstanding amount, and proceedings were again being taken against them in September 1857, when Perks reported that in addition to previous debts they owed £261 14s. in respect of tonnages to 30 June. In January 1858, he showed the meeting a bill of exchange for this amount, accepted by the Talybont Coal Co., which was due later in the month, and it was decided that should it be unpaid on presentation, action should be taken without waiting for the management's authority. Such was the credit worthiness of the Talybont Coal Co., and in fact, in the same month the B&A refused them any further credit, and ordered that their tonnages should be paid before the boats were allowed to move.[36] By May 1858, the coal company had taken up the tramroad connection with the Brinore and removed the line along the canal bank, originally laid by Powell & Co., very much to the disgust of Mary Dixon, who lost her perquisite of the sweepings from the kilns with the cessation of the limestone traffic. A special meeting on 31 May expressed sympathy but was unable to suggest a way of helping her. The Brinore also had difficulties, both by loss of the traffic, and by having to find accommodation at their own wharves for the coal company's traffic in coal.

Perks' report of 3 January 1859 revealed that the Talybont Coal Co. was known to be in difficulties, and as they owed £216 16s. 9d. in respect of tonnages to the end of the previous September he had distrained on the trams, coal and lime which had been lying on the tramroad company's property. They had paid off £90 8s. of the sum due, and he had accordingly released their trams and coal, but he had retained the plates lifted from their branch tramroad as security for the remainder of the debt. At the end of the month Perks was authorised to purchase the plates at £6 per ton for the use of the Brinore, applying the proceeds to the reduction of the tonnages due, and to accept a bill for the balance then owing, £61 16s. 9d. With regard to the debt of £261 14s. due for tonnages up to 30 June 1857, it was agreed that the company should accept a bill for £100, guaranteed by David Thomas, the solicitor, in part payment.

Perks informed the meeting of January 1860 that he had been unable to obtain payment of tonnages amounting to £108 8s. 1d. due the previous June, and in November of the same year he reported that there was a further amount of £209 15s. 9d. due for tonnages up to June 1860. The Talybont Coal Co. disappears from the minute book after a despairing effort made on 5 November to obtain payment of the latter amount.

The exact date when Overton & Scroop went out of business has not been determined, but their account with the Rhymney Iron Co., from whom their lease of Brynoer Colliery was then held, was closed in July 1842,[36] and it is assumed that they then ceased to trade. It is probable that their lease was taken over by Powell & Co.

The only other freighter on the tramroad to be traced was a firm by the name of Lloyd & Watkins, which is mentioned briefly in the accounts in 1858, when they were written off as a bad debt in the sum of £13 16s. 7d.

Through the years various railway schemes were proposed, designed to supercede the B&A, none of which would have had an adverse effect on the Brinore, as all were on a direct line from Abergavenny to Brecon. The proposal for the Brecon & Merthyr Tydfil Junction Railway in 1858 was, however, a very different matter as far as the tramroad company was concerned, as the deposited plans showed part of the tramroad within the "Limit of Deviation".

In a newspaper report of one of the promotional meetings[37] " a gentleman of Rhymney" (possibly William Llewellin) suggested he could not understand why a route from Rhymney to Brecon via the Dyffryn Valley had never been considered, and at another meeting held in Merthyr Tydfil, George Overton suggested that a line from Sirhowy to Brecon might be built as a branch, utilising the line of the tramroad.[38] On 26 November 1858, Rev. T. J. Powell wrote to G. T. Clark, of the Dowlais Iron Co., who took a great interest in the promotion of the B&M, suggesting that it would be a gross injustice if they were to be permitted to lay a line of railway parallel to the Brinore without giving adequate compensation, in view of the manner in which the tramroad had come to the aid of the town of Brecon when that community had been in imminent danger of being victimised by a monopoly in the supply of coal and lime. He suggested that should the B&M see fit to purchase the shares of the Brinore, he had no doubt that the money would be invested in a branch railway from Rhymney to Talybont, thus providing a feeder from Rhymney Ironworks to the B&M. He offered to call a general meeting of the tramroad proprietors to discuss the matter. His postscript probably gives the true reason for his letter, pointing out that the B&M had not yet appointed anyone in Brecon to look after their interests, and suggested that his son, Gabriel William Powell, who was a solicitor in the town, might be a suitable candidate for the position.[39]

A meeting of the Brinore on 3 January considered the threat that the railway proposals presented, and appointed a special committee to go into the matter of obtaining siding accommodation at Talybont for their use, and to consider on what terms the company could offer its undertaking to either the B&M or the Breconshire Railway & Canal Co.,* and to report on both matters to a special general meeting to be held on 31 January 1859. The later meeting decided that Thomas should write to J. R. Cobb, who had been appointed solicitor to the B&M in Brecon, offering them the tramroad concern at £60 per £100 share for cash or £80 per share if paid in preference shares at five per cent, the choice to be that of the railway company. Thomas was also instructed to prepare petitions against the Bills of both railway companies.

How little the B&M thought of the opposition of the Brinore was revealed in a letter from their secretary to Cobb on 2 February in which he remarked:

> "As for the Brynoer from what I hear I doubt if their whole year's revenue would enable them to appear before Committee [of the House of Lords]. I learn that on Monday Mr. D. Thomas tried hard for powers to oppose the canal line and the Merthyr one there on their behalf but only received instructions to *prepare* petitions".[40]

Nevertheless, on 2 February the Brinore decided to proceed with both petitions. Cobb, who was in London watching the progress of the Bill, wrote to Clark on 10 March, " . . . as to the Brinoer matter I think we cannot do better than put it wholly in the hands of our Parliamentary Agent".[41]

The B&M Bill received Royal Assent on 4 July 1859, and the *Cambrian* of 9 July reported that the Brinore petition had been rejected in the Lords, where it had been ruled that they had no right to be heard, but their Lordships had granted them an extra three feet headroom where the B&M crossed their line. In the event, the line on which the railway was built only crossed the Brinore at a point where the branch made by Powell & Co. had been, so no inconvenience was caused.

In general, the Brinore shareholders seem to have accepted the position philosophically, with the exception of Rev. T. J. Powell, who wrote to Llewellin in November 1859 suggesting that he should contact Hall, by then Lord Llanover, and recently President of the Board of Trade, in an endeavour to obtain some form of compensation for the proprietors of the Brinore.[42] Llewellin's relpy has not been found, but it is doubtful if Lord Llanover could have done anything to wring charity from the B&M. Conybeare, the engineer of the B&M, reported to his shareholders that he thought the " . . . Brecon Canal and the Brynore Tramroad, by the convenience they afford for the conveyance of materials will greatly facilitate the construction . . . ",[43] and that was probably the view of most of those connected with the railway—the tramroad era was over as far as Breconshire was concerned.

V

In December 1860, there was a change in the usual proceedure; instead of the Brinore approaching a prospective purchaser, it was itself approached by the proposed Sirhowy & Breconshire Railway with the suggestion that it should sell itself to that concern for £40 cash and

* A joint scheme of the B&A and the Newport, Abergavenny & Hereford Railway, which failed to obtain its Act.

one £10 share in the new company for each £100 tramroad share. Almost immediately the S&B increased its offer to £50 cash per share, which was too good to be missed in view of the low state of the tramroad finances, and Thomas was instructed to prepare a provisional contract for the consideration of E. J. C. Davies of Crickhowell, the S&B solicitor, and he was to give him an assurance that, provided a clause was inserted into the S&B Bill covering the purchase, no petition would be lodged by the Brinore against their scheme. A special general meeting was called, to be held on the same day as the ordinary meeting on 8 January 1861, to confirm the resolution.

The general meeting was duly held, and after conducting the normal business turned itself into the special meeting. One can imagine the consternation when Thomas read a letter he had received from Davies varying the terms on which the S&B was prepared to acquire the Brinore to a substitution of a rent charge of five per cent per annum on the proposed cash purchase price, in lieu of full cash payment. As it had been convened for the sole purpose of confirming the original proposals, the special meeting did not feel itself competent to consider the new proposal, and promptly reverted back to an ordinary meeting, which could entertain new business.

Davies' letter stated that he was not empowered to deviate from the terms stated, but in view of the altered situation the Brinore set about imposing its own stipulations, laying down that the rent charge should become payable on the passing of the S&B Act; that the S&B should not touch any part of the Brinore property until such time as the S&B line from the B&M at Talybont to the Rock Inn, Penrhiwcalch, and from Trevil, through the intended tunnel, to the head of the Crawnon Valley should be ready to be opened to traffic, and that the Brinore should continue to manage its own affairs until Royal Assent had been given to the S&B Act and after that time the net receipts of the tramroad were to be paid to the railway in return for the rent charge.

Another special meeting was held on 22 January, when another letter from Davies was read suggesting further modifications to the agreement which would have permitted the S&B to take possession immediately on the provision of a suitable guarantee for the payment of the rent, and that the railway should have the use of the tramroad, free of charge, for all purposes in connection with the railway works. The Brinore did not like the new suggestions particularly, and adjourned the meeting to 26 February to enable further discussions to take place.

The meeting of 26 February decided to delay making any decision as the S&B Bill was that day before the Standing Order Committee of the House of Commons, and was adjourned until the 28th, when the result of the committee's investigations would be known. The news of the Bill's rejection, due to failure to pay the deposit, was received with apparent equanimity and it was decided that it was useless to pursue the matter any further. Thus ended what proved to be the last opportunity for the Brinore to sell itself to a railway company.

Sufficient traffic passed in the year ended 30 June 1861 for a dividend of three per cent to be declared, but in the next year traffic declined, and two per cent was all that could be paid. Perks' report of 14 October 1862 makes melancholy reading with receipts down approximately £100 on the previous year. With the track "quite equal to the present traffic" he had not considered it wise to do any new work due to the approaching opening of the B&M, and the only maintenance he had done was to straighten and re-ballast the worst parts.

Belatedly, in January 1863, the tramroad company had doubts as to the legality of its attempts to dispose of itself without the consent of the B&A, under whose Act it had been built, and decided to obtain counsel's opinion on this and if it had the right to determine its own tonnages. The opinion was received in time to be reported to the May meeting. Whilst it is not detailed in the minutes, it is evident from future events that it was considered in order for them to dispose of their property, but that the tonnages were governed by those currently payable on the B&A.

With its finances in a bad state it was unfortunate that the question of tonnages could not be side-stepped, and a resolution was passed reducing them to those current on the B&A, to take effect retrospectively to the previous January, when the B&A had made a reduction.

The same meeting reduced Perks' salary to £52 per annum as an economy measure, and in January 1864, it was decided to dispense with the services of the clerk at Talybont whose duties were to be undertaken by Jenkin Jenkins, "the workman employed on the road", under the supervision of Perks, and Perks himself suffered a further reduction in salary to £32 per annum, "in consequence of lessened duty". Perks had his own ideas for reducing expenditure, and his suggestion that the rails from the now virtually useless passing places should be taken up and used to repair other parts of the track was agreed to. The immediate cause of the company's difficulties was the expenses incurred in their opposition to the B&M, and after paying Thomas "£100 further on account" of his bill of costs for the exercise, the balance remaining in the possession of the company amounted to only £60 6s. 5d. From this date, dividends became a matter of history, and it is also the last reference to T. C. Perks who disappears from the records, and his further activities are unknown.

There is no record of the proceedings of the company for the next seven years, the next recorded meeting taking place on 6 November 1871, at the office of J. R. Cobb, who was now taking an interest in the company. With Cobb in the chair, it was decided not to disband the company, but to convert the tramroad into a stone road, and to sell the track parts. It was, however, resolved that no action should be taken in this direction until all members had been circularised notifying them of the intended action, and the meeting was adjourned until 4 December.

At the meeting of 4 December, Cobb stated that he had notified all *known proprietors* and that one letter only had been returned through the dead-letter office, and he proceeded to move that the tramroad be altered to a stone road, but that no action should be taken until the following 25 March in case an offer was received for the concern as it stood. An amendment moved by Llewellin and seconded by Overton that no action should be taken was defeated, and the original motion was passed. John Tudor was appointed Clerk to the concern, without salary, and he was instructed to arrange with Jenkin Jenkins to take tonnages for such traffic as was presented. This is the only evidence that has been found that traffic was still passing on the tramroad, and is the last mention of receipts.

By September 1872, it had been suggested that it might be *ultra vires* for the company to remove the track, and Thomas was instructed to obtain counsel's opinion on the point and the meeting held on the 30th of that month was adjourned without discussing the proposition, in the order calling the meeting, that the company should register under the Joint Stock Companies Act of 1862. The next meeting, on 28 October was informed by Thomas that he had not yet applied for the required opinion, as no resolution had been passed naming to whom he should apply. The Meeting passed four resolutions:

1. That Thomas should apply to Mr. Southgate, Q.C. or failing him to Mr. Cotton, Q.C. for the requisite opinion.
2. That the previous resolution to convert the tramroad into a stone road be acted upon, and the rails sold.
3. That the company should be registered under the Joint Stock Companies Acts of 1862 and 1867, and immediately wound up.
4. Should counsel's opinion require it, the company should apply for an Act of Abandonment.

A fortnight later Thomas read out Mr. Cotton's opinion that the company could not convert the tramroad, nor sell the track without the prior sanction of Parliament, and reported that he

Brinore Tramrod: Penrhiwcalch Bridge, demolished in 1958 and photographed in 1947.

had prepared the notices necessary for an application for the Act to abandon the concern. It was decided that the application should be delayed until the following year in the hope that in the meantime a purchaser for the tramroad might be found, and that if an offer of £6,000 should be received all present would concur in a sale at that price.

It took until March 1874 for the Bill to abandon to be prepared, but only two months further for it to be rejected by the Commons. The minutes do not state the reason for the failure, but merely the fact of the failure and the resolve to apply again in the next Session. The new Bill was read and approved by the proprietors at the meeting of 17 February 1875. It is known that this failed, but there is no mention of the fact in the minutes.

The failure of the Bills may be attributed to the Brinore having been constructed under the B&A Act, and when that canal was taken over by the Monmouthshire Railway & Canal Co. in 1865, the Act for the amalgamation specifically forbade the closure of any part of the B&A or its tramroads without Parliamentary authority; and the House probably considered them to have no *locus standi* (as had happened with their petition against the B&M), and that to be effective the application should have originated with the MRCC.

Only one more meeting of the tramroad company is recorded. On 24 February 1890, a meeting convened by J. R. Cobb considered an offer he had received for the track. He stated that he had written to or seen all the shareholders whose addresses he knew, and those known to be the representatives of deceased proprietors, but nonetheless there were present only J. R. Cobb, C. E. W. Cobb, and John Tudor, and two others represented by proxies. At J. R. Cobb's request,

John Tudor read out the minutes of the meetings at which it had been resolved to apply for an Act of Abandonment, and Tudor stated that as a result of those applications expenses had been incurred, which he, as Clerk, had no funds to meet, and that he had been served with process for £690 13s. 7d. on account of Parliamentary expenses. J. R. Cobb confirmed that, as solicitor to the company, he had made the disbursements, and offered to waive his professional fees if the processes were met. The company having no money, no instructions were made for an appearance to be made to the writ.

It was reported that an offer of £1,200 had been received for the plates, at £2 per ton, but as this was 10s. per ton less than the same person had offered only a month previously, the offer was turned down.

The meeting was adjourned to Monday, 31 March 1890, but no record survives to show if this was held, and there are no further entries in the minute book.

Thus, like an old soldier, the Brinore faded slowly away, but even today, its ghost occasionally walks. In 1958, Penrhiwcalch Bridge was judged to be unsafe, and after making extensive efforts to trace an owner, the late Breconshire County Council obtained a compulsory purchase order for it, and the cutting spanned by the bridge was filled in where necessary to provide a stable road, and a small sum deposited with a bank in Brecon, where doubtless it will lie unclaimed.

According to the published accounts for 1857 and 1858, the company owned some property, as there are entries for rents received, probably for the cottages lying on the line of the tramroad above the canal at Talybont (SO.1150 2249). About 100 yards away, at SO.1140 2246, is the weigh-house falling into ruin, whilst the office of the company at SO.1178 2229 is in a state of advanced decay. The company owned, and possibly still does, the land over which the tramroad was carried, and over which there is still a way open to the public. Many indications of its former use survive, but there is now no-one to collect a toll, not even the shade of Jenkin Jenkins.

VII Trevil to Brynoer Colliery and Rhymney

This section of the line, although a continuation of the Brinore, was built privately at the expense of Benjamin Hall, and whilst it was open to the public, was used mainly for the traffic to and from the Brynoer Colliery and the ironworks at Rhymney.

Evidence that it was a 'public' tramroad is found in the Brinore Tramroad minute book, where under the date 21 February 1814, the solicitor, Samuel Church, was requested to obtain Hall's guarantee that the continuation of the line, to be made by him, would be " . . . secured as a Public Road . . . ", and evidence of its construction comes from the same source on 25 April 1815, when John Llewellin, who probably had charge of the operation, reported to the Brinore shareholders that only 200 yards remained to be completed, and there can be little doubt that it was ready for traffic in time to be opened with the Brinore.

The lease of the Brynoer Colliery to Dixon & Overton of 29 July 1814 only gives the location vaguely as " . . . mine veins and seams of coal lying under sixty acres of Bedwelty . . . " adding that it was then in the occupation of Rosser Thomas.[44] An assignment of his lease by Thomas dated 1816 describes his holding more definitely as bounded on the east by Ras Brynoer (SO.116 091), and on the west by the tramway leading from Brynoer Colliery to the Upper Furnace, describing the locality generally as Ffos yr Hebog (SO.116 088).[45] The 1st edition O.S. 1″ shows Brynoer Colliery at SO.1168 0912, whence a tramroad is shown connecting with the line to Trevil at Tavarnaubach. On the 1st edition O.S. 6″, "Brynoer Level" is shown at SO.1136 0886, with a tramroad joining the line to Trevil at SO.1105 0896, now in the middle of a housing estate. The books of the Rhymney Iron Co.[46] base all tonnages charged from the colliery to the terminus of the Brinore at the "eight mile mark" on a distance of three and a quarter miles, and measurements taken on the modern 6″ O.S. show that this distance from the location shown on the 1st edition O.S. 1″ map via the tramroad coincided with the position of the eight mile mark, as measured over the line of the Brinore. If the distance is measured on the line shown on the 1st edition O.S.6″ map, it places the end of the Brinore proper about half a mile nearer to Talybont. It would, therefore, appear that, at the time in question, the O.S. 1″ map gives the correct location.

In early days, before the amalgamation of the Union Co. with the Bute Iron Co. in 1835, large quantities of limestone were probably carried to both the Upper and Lower Furnaces, but after that date the quarries of the Bute Co. at Twynnau Gwynnion (SO.065 113), near Pontsticill would have been available to the joint concern. By 1852, the lease of Trevil Quarry had been given up, and a renewal in that year refers only to the land over which the tramroad had been built by Hall, and in fact the lease specifically lays down that only "Pitwood, Hay, Oats, Straw and Coal" may be carried.[47]

Even as late as 1862, fairly large quantities of pitwood are shown as being carried over the Brinore. As there are no collieries to the north of Tavarnaubach in which these could have been used, it follows that they must have been carried over Hall's line to Brynoer Colliery or to be used at one of the collieries supplying the ironworks. No record has been found of the carriage of iron on the Brinore, but doubtless, in early days before the opening of the Old Rumney *circa* 1828, Hall used this route to take his iron to the B&A for onward carriage to Newport. In later years, the sole function of his tramroad and the Brinore as far as the ironworks was concerned must have been to carry pitwood and possibly a certain quantity of agricultural goods such as the Rhymney Iron Co. was unable to produce on its own farm at Rhymney.

The volume of limestone traffic can be judged to a certain extent from the inventory of the property of the Union Co. at the amalgamation.[48] This shows that there were ninety-two trams assigned to this traffic, each valued at £2 10s., which shows that they must have been considerably smaller than those used on the Old Rumney which were valued at £6 10s. each. The latter for use on a four feet two inch gauge, probably carried three to three and a half tons of iron; the former for a three feet six inch gauge can only have carried between one and one and a half tons.

A somewhat puzzling entry in the inventory is the following:

"Tramplates and Sleepers. 239T. 7Cwt. 0qrs. 121lbs. Overton's Triple Road. £1,316. 9. 1.
Tramplates and Sleepers. 82T. 3Cwt. 0qrs. 12lbs. Overton's Double Road. £451.17.2."

It is difficult to determine to what these entries refer. There were three different gauges used in the vicinity of the works—the iron company's internal gauge of three feet, the Brinore line of three feet six inches, and the Old Rumney at four feet two inches. Does the "triple road" refer to a mixture of all three? Was it a "gauntleting" of any two? Or was it just a matter of three parallel tracks?

The extent to which Hall's line was used by the public can be judged by the account of Giles Jones, who was his toll-collector at Trevil. The following figures are all for years ending 30 June.[49]

1837.	£12.15s. 1d.	1840.	£10. 3s. 4d.	1843.	£39. 0s. 1d.
1838.	£13. 6s. 3d.	1841.	£12. 1s. 1d.	1844.	£21.14s. 8d.
1839.	£17. 2s. 5d.	1842.	£15.10s. 6d.	1845.	£24.12s. 2d.

The traffic from Dixon & Overton's colliery does not show in the above figures, as this was paid direct to the company with the royalty on the coal raised, nor does it account for limestone sent to the works from Trevil Quarries. The following figures for the tonnages of Dixon & Overton have been calculated from the gross royalty and tonnage figures shown in the ledgers, and it has been assumed that there was no variation in the terms of the original lease:

1837. £160.11s. 9d. 1838. £164.11s. 5d. 1839. £169. 1s. 8d. 1840. £172.10s. 5d.
1841 (six months only). £68.12s. 7d.[50]

It appears that Overton's lease was transferred to the Talybont Coal Co. as from January 1841, and although the figures in their account are similar to those appearing under the heading of 'Overton & Co.', no attempt has been made to calculate the amount of their tonnages, as it is not known if the terms of their lease were the same as those of the previous tenants. Figures produced by Giles Jones for the six months from May to October 1850 show that he weighed 2,894 trams containing 3,595 tons of coal in the period, but he does not identify the consignor.[51]

A re-routing of the pitwood traffic undoubtedly took place in 1868, when the B&M line from Dowlais Top to Bargoed was opened.[52] Their line had a junction at SO.0864 0845 with the Rhymney Iron Co's. limestone railway running from the Bute Works to Twynnau Gwynnion quarries, and thus provided access for rail-borne traffic directly into the works. It is unlikely that there was any appreciable traffic on Hall's line after this date.

VII Miscellanea

Whilst Talybont was the administrative centre of the Brinore, it could possibly have been more important had a scheme of George Overton, Senior's been put into effect. In 1821 he prepared a plan to construct a tramroad, four miles long, from Talybont to connect with the Hay Railway at Talyllyn[53] which, had it been made, would have considerably reduced the tonnages on cargoes carried to the north of that place by obviating seven and a quarter miles of canal and three and three-quarter miles of the Hay and substituting four miles of new tramroad, which on coal would have effected a saving of 1s. 2d. per ton at Talyllyn. In his book, published in 1825,[54] Overton blames the non-fulfilment of his scheme on opposition from one landowner and the B&A. Indeed, the possibility of such an extension had obviously been discussed as early as 1816, as in that year the B&A took counsel's opinion as to their rights should it be made, and if they could, under their Act, legitimately take more than the minimum tonnage of a quarter of a mile on goods carried over the canal,[55] which was ruled to be the most they could expect.

To ascertain the type of tram used on the Brinore it is necessary to turn again to Overton's book[56] in which he describes trams that he had designed some twenty years before as:

> "... constructed with sides one foot upright and nine inches upon that to bevel a little outwards ... [he then gives the dimensions of the wood to be used in the construction] ... This description of tram I introduced more than twenty years ago and it has since been adopted at most of the works throughout South Wales. I have ever since adhered to the same shape and form, but have substituted iron instead of wood, using gunnel iron for the sills ... flat iron for the straps, ... and plate iron one eighth of an inch thick for the sides and bottom. Ten years ago the Penydarren Company made me two hundred trams of this sort in one lot; and although somewhat worn, they are still as good for every purpose of practical use as they were at first".

Obviously, Overton was not a very modest man, but we must allow him his moment of triumph! As the above was written in 1825, one can assume that the iron trams were intended for use on the Brinore, opened in 1815.

The number of trams in this order gives some idea of the quantity of coal that it was expected would be carried from Brynoer Colliery, but it is not until 1836 that it is possible to assess the actual amount of coal carried, when the books of the Rhymney Iron Co. show that in the years 1836–41 an average of 8,500 tons annually passed over Hall's line.[57] This quantity would have been insufficient to meet the 'dead rent' on the colliery, where the lease called for a minimum of 9,230 tons annually. It was, in fact, only sufficient to meet the minimum daily requirement of twenty-five tons, assuming a 300 day working year. In only one year, however, do Overton & Co. appear to have been charged with a sum to make up the dead rent, when in 1837 they were charged £111 14s. 7d. in addition to their royalty payment.[58]

We know that from 1820, William Powell & Co. were trading on the Brinore. By 1825, it would appear that they were expecting their business to expand considerably when they leased land at Trevil from the Duke of Beaufort, on which they are reputed to have built three cottages, an inn and stables,[59] but by 1829 the lease was assigned to Giles Jones, in whose family it remained until 1855 or later.[60] In view of the location of the premises, which are stated to be between the Brinore Tramroad and the Trevil Brook, it must be assumed that this was the same Giles Jones who appears as toll collector in the Iron Co's. books in the late 1830s and early 1840s.

VIII Route

The most obvious reminder of the tramroad at Talybont is the bank of kilns at SO.1168 2240, to the top of which the tramroad ran, and on the canal side can be seen the widening effected by Jonathan Dixon. A line to the canal came down to SO.1158 2251, and this probably marks the point of shipment for coal. This line rises sharply to meet the kiln line at SO.1149 2249, where are the cottages previously mentioned.

The office was at SO.1175 2235, some fifty yards from the kilns, and comprises a small two storied building with lancet windows set in iron frames. The upper storey is level with the road from Talybont to Penrhiwcalch, and the lower on the level of the canal bank. There is a triangular piece of land between the road and the canal, and it is thought possible that this was the garden allocated for the use of the various clerks of the company.

William Powell & Co's. wharf was probably at SO.1145 2251, adjacent to the canal waste weir, but all traces have vanished, being covered in the main when the B&M was constructed. Their branch tramroad has disappeared from the same cause, but would have met the Brinore slightly to the north of the weigh-house at SO.1140 2245. This is still standing but is decaying rapidly.

From this point the tramroad can be followed easily, climbing steadily in a southerly direction, and for the first mile or so the hedges are still maintained, as the tramroad provides access to the fields belonging to the farm at SO.1079 2168. At SO.1086 2012, however, it enters the coniferous plantation of the Forestry Commission, and only the passage of walkers keeps the path open. There are sleeper blocks on this stretch, and in several places the track widens indicating the sites of passing places.

Along this stretch are the ruins of several buildings, the purpose of which is not known, although it is reputed that one was an illicit beer-house, the owner of which regularly paid fines for serving alcoholic refreshments to the hauliers.

At Penrhiwcalch (SO.1012 1765), the tramroad crosses from the Caerfanell valley, in which is the Talybont Reservoir, over the col into the Crawnon valley, by a short shallow cutting. Until 1958 this was spanned by a small stone bridge carrying the Roman road from Talybont to

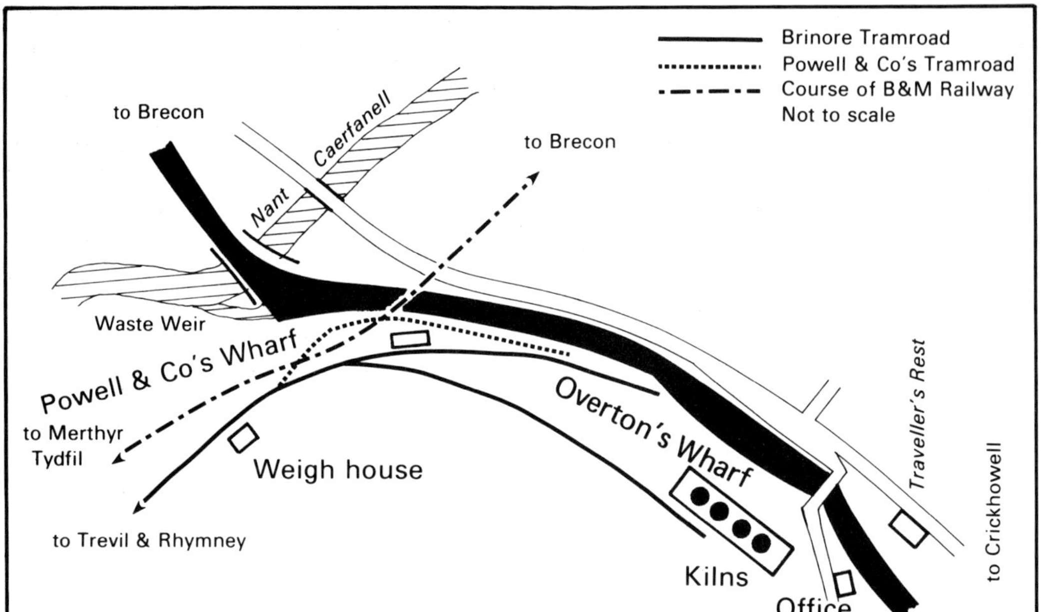

Talybont on Usk

Pontsticill and Merthyr Tydfil. In that year it was adjudged unsafe even for the sparse traffic it had to carry, and an embankment was made over the cutting.

As its name implies, Penrhiwcalch (Head of the lime hill) was the place to which, in pre-tramroad days, the farmers resorted to collect lime for use in the Caerfanell and Crawnon valleys. With the advent of the tramroad, and the building of kilns on the canal side, this trade was probably lost, unless it was done on a 'return load' basis. That there was a considerable trade of some sort here is shown by the existence, at the southern end ofthe cutting, of a number of sidings, with one from the end of the cutting leading up to the level of the Roman road.

A ruined building at SO.1015 1765 was probably the 'Rock Inn',[61] and trade at this place was sufficient, even as late as 1841, for the tramroad company to erect " . . . a Shed . . . for the Benefit of the Traders on the said Road" the cost of which was not to exceed £32. At that price a fair sized building could have been erected, and it was probably in the nature of a warehouse in which freighters could leave goods to be collected by consignees. In 1857, £2 8s. 6d. was received for its use, and in 1858, £12 16s.[62] The location of the 'Shed' has not been ascertained.

Continuing southward, on the west of the tramroad, about 100 yards from the cutting, there is a ruin, which appears to have been a dwelling-house. For the next half-mile, the line is defined by hawthorn hedges which bear evidence of having been layered, and are of considerable age and are possibly original plantings.

At SO.0962 1678, the line enters recent coniferous planting by the Forestry Commission, and whilst there are stone blocks in plenty to prove the former usage of the track, it is, unfortunately, now very enclosed and damp under foot, as the Commission's activities have upset the drainage which twenty-five years ago was still in good condition.

At SO.0930 1573 the tramroad comes to the steep hillside at the head of the Crawnon Valley and the track is either on a side-cut, or embankments to cross small side valleys. At SO.0907 1560 the side of the valley is so steep that the Commission has been unable to plant it, and one looks over the tops of the nearest trees, planted far below.

There must have been an exceedingly sharp turn onto the embankment that crossed the stream at Craig Ddu (SO.0907 1553), but this has now entirely disappeared, as has the arch of the culvert at the base of it. This was partly discernible as recently as 1964.

For the next half mile, the tramroad has been cut into the almost vertical end of the valley, and at the time it was made must have been a major feat of engineering, as only black powder would have been available for blasting. Unfortunately, the valley faces due north, and suffers the worst of all the winter weather with the inevitable result that erosion has left its mark in no uncertain manner, and the ledge which twenty years ago was a pleasant walk along a path some eight feet wide is now a scramble along a narrow path sometimes less than two feet wide. In places this is still crumbling, and the fence made by the Forestry Commission has the posts hanging by the wires.

There were some very old kilns at SO.0927 1113, but on the writer's last visit in May 1976 these had been hidden by a rock fall. The reference is only quoted in the hope that further earth movements may reveal them again.

At SO.0942 1500, where the old road up the valley climbs to join the line of the tramroad, the line is built on more solid ground, as a natural ledge exists which the tramroad builders used, and so level is the strata that in a number of places the cast iron ties were laid directly on the bed-rock; some places have been found where the level was not quite good enough, and small pieces of broken plates inserted as packing have left their mark on the rock.

After passing through a shallow cutting, the tramroad crosses the Afon Crawnon by an arch built right on the edge of a thirty-foot cliff, down which the river falls, in flood times being quite spectacular (SO.0953 1492). It then passes old quarry workings, which due to weather and vibration from recent nearby quarry workings have shed rocks all over the tramroad.

The kilns at SO.0992 1506 were probably worked by Dixon & Overton in connection with their adjacent Crawnon Quarry, and from here the tramroad looks more like a cart-road than a tramroad, having at some time received a top-dressing of stone.

At SO.1083 1523 the tramroad crosses the line of the railway constructed in 1938 by Richard Thomas & Co. Ltd. in connection with the Ebbw Vale Ironworks, and the route of the Brinore becomes a tar-macadamed road. Until the early 1950s the tramroad blocks could be seen on the west side of the road, which was then unsurfaced. From SO.1113 1500, at Ffos y Wern the road descends steadily to Trevil.

The exact site of the "Eight Mile Mark" has not been found, but measurement on the 6″ O.S. shows that it must have been at about SO.1189 1308, about a quarter of a mile north of the *Quarryman's Arms* public house, Trevil.

Hall's tramroad continued on the line of the road, crossing the embryo Sirhowy river at SO.1201 1205. At SO.1271 1143, it diverged to the west passing through Penrhyn Farm at SO.1266 1131, the hedge-line leaving the road at an acute angle marking its line.

Passing through the farm the tramroad turned to run south-westward to SO.1269 1090, where it is now covered by an industrial estate. Emerging at SO.1209 1054, it passed in front of the chapel at SO.1205 1039 and behind that at SO.1201 1030, and behind the row of houses at SO.1191 1025. The hall at the end of the row of houses at SO.1188 1025 is built on the tramroad site.

Crossing the old Merthyr road, the tramroad was carried over the MTA by a skew-bridge at

SO.1175 1017. The line of the railway at this point was used to make the Heads of the Valleys Road, and to the south of it the whole area comprising Bryn Brith, Brynoer and Bryn Bach has now ben open-cast removing all landmarks to which the tramroad might be related. The map references in the next paragraph have consequently been taken from the writer's observations made before opencasting took place and corroborated by old maps.

At SO.1175 1005 the line divided, a branch to the east of the main line running straight to Brynoer Colliery at SO.1168 0912; this branch was undoubtedly made by Hall to handle Dixon & Overton's traffic. The main line ran south-westerly, passing to the south of Bryn Brith Pond, and was joined at SO.1128 0921 by a line from Bryn Brith Colliery (SO.1132 0955).

It continued south-westward to enter the site of the present housing estate at SO.1105 0895, and curved to the south to join the main road (A469) at SO.1106 0868. The 1″ O.S. of 1831 shows it as running from this point to kilns at approximately SO.111 081, and there was a tramroad running alongside the river Rhymney to convey the dried stone to Upper Furnace at SO.1085 0916 and Lower Furnace at approximately SO.111 079.[63] A plan of Rhymney in 1846 shows the ironworks to have been moved to the south at SO.114 072, and the sites of the kilns and Lower Furnace are shown as "Works buildings and Rubbish".[64]

The 6″ O.S. survey of 1873 (published in 1884) shows the tramroad as running on the site of the present main road from SO.1106 0868 to enter Church Street at SO.1102 0814, where it is shown to have been on the site of the buildings to the east of the road. In front of the Castle Hotel (SO.1130 0795) it was joined by an incline from Gwaunfawr Pit (SO.1149 0841) and continued down the existing lane to the rear of the buildings in High Street.

Following the lane, it crossed Victoria Road at SO.1149 0755 and can be followed for about another 150 yards where it is lost in the industrial desolation that was Rhymney Ironworks.

The Brinore Tramroad was laid on cast iron plates held to gauge by cast tie-bars forty-nine and three-quarter inches long overall, with chairs for the rails cast integrally; with the ends of the ties resting on stone blocks, and held in place by the ballast.

The forty-seven inch long plates were held in the chairs by wooden keys hammered in on the flange side, giving a back to back of the plates measurement of thirty-nine and a quarter inches, and a face to face measurement of forty-one and a quarter inches, this allowed three-quarters of an inch side-play to the wheels of the trams, which were set to the nominal gauge of the line, forty-two inches.

With the use of tie-bars, the stone blocks had to be maintained opposite one another, and to achieve this, special plates with a length of forty-seven and a half inches were used for the outer rail of curves.

Turnouts appear to have been of the normal plateway type, spiked to stone blocks. To effect a connection between the two types of track special plates were used, with one end designed to fit the tie-bars and the other slotted to take the spike.

Such scant remains as have been found on Hall's line indicate that this was laid with the normal type of spiked track, with plates thirty-six inches long and weighing forty-five to fifty pounds.

No evidence had been found of the general use of wrought iron plates, but one was found lying some ten feet below the tramroad level at the head of Dyffryn Crawnon. This is the sole example that has been seen, but scrap drives in two world wars may have accounted for any others that there might have been.

REFERENCES:

Where quotations have been made and no reference is stated, these have been taken from the Brinore Tramroad Company's minute book, which has been copied by courtesy of the Breconshire Railway Society.

1. Edwards, E. E. *Echoes of Rhymney.* Rhymney U.D.C. 1974.
2. PRO. B&A proprietors minutes, 15 October 1812.
3. PRO. B&A committee minutes, 3 December 1812.
4. NLOW. Maybery 918.
5. GRO. Man/G/1-0006.
6. NLOW. Maybery 923.
7. ibid. 924.
8. *Cambrian*, 6 February 1813.
9. ibid. 13 March 1813.
10. PRO. B&A committee minutes, 29 April 1813.
11. *Cambrian*, 23 April 1813.
12. At the time, anglicised spelling of Welsh names was common. These were usually phonetic.
13. NLOW. Maybery 273.
14. ibid. 269.
15. GRO. D.10/1382, and D.10/1663-1672.
16. ibid. D.10/1661.
17. Brecknock Museum, Brecon. Maybery collection. Vol. 2. fo. 116.
18. NLOW. Maybery 973.
19. *Herford Journal*, 26 July 1820.
20. NLOW. Maybery 2113-4, 2395 and others.
21. PRO. B&A committee minutes, 27 April 1820.
22. ibid. 1 February 1821.
23. PRO. B&A proprietors minutes, 26 April 1821.
24. GRO. D.43/1322.
25. R. S. Harford paid £87 for the share. Brinore Tramroad minutes, flyleaf.
26. PRO. MCC committee minutes, 23 June 1843. The MCC solicitor was ordered to attend a meeting of creditors of Harford, Davies & Co. at Bristol on 27 June.
27. *Hereford Journal*, 9 February 1825.
28. ibid. 26 January 1820.
29. ibid. 20 December 1825.
30. T. P. Price was a partner of W. Bridgewater in the Talybont Coal Co. (Brinore Tramroad minutes, 31 May 1858.)
31. GRO. Man/G/1-0006.
32. Barrie, D. S. M. *The Rhymney Railway.* Oakwood Press, 1963.
33. GRO. Man/G/1-0006.
34. Clinker, C. R. *The Hay Railway.* D&C. 1960.
35. PRO. B&A committee minutes, 28 January 1858.
36. Glamorgan Record Office. D/Rh.53.
37. *Cardiff & Merthyr Guardian*, 2 October 1858.
38. ibid. 20 November 1858.
39. NLOW. Maybery 1038.
40. ibid. 2413.
41. ibid. 2417.
42. GRO. Man/G/1-0006.
43. *Cardiff & Merthyr Guardian*, 9 March 1861.
44. NLOW. Maybery 269.
45. GRO. D.43/825.
46. Glamorgan Record Office, D./Rh.12, and 53.
47. NLOW. Badminton 6092.
48. Glamorgan Record Office, D/Rh.14. fo. 2.
49. Glamorgan Record Office. Various works ledgers under ref. D/Rh.
50. ibid.
51. GRO. Man/G/1-0006.
52. Barrie, D. S. M. *The Brecon & Merthyr Railway.* Oakwood Press, 1957.
53. Brecknock Museum, Brecon. A.79/2/3(57). Plan and estimate.
54. Overton, George. *Description of the Faults & Dykes of the South Wales Mineral Basin.* 1825. Vol. 1. Vol. 2. does not appear to have been written.
55. NLOW. Maybery 973.
56. Overton. op. cit.
57. Glamorgan Record Office, D/Rh.12 and 53.
58. Glamorgan Record Office, D/Rh.4.
59. Davies, Emrys. *The Brinore Tramroad* in *Presenting Monmouthshire* (Journal of the Monmouthshire Local History Council), No. 26. Autumn 1968.
60. NLOW. Badminton 5959-60.
61. Brecknock Museum, Brecon. Maybery collection, Vol. 1A. fo. 65.
62. GRO. Man/G/1-0006.
63. Glamorgan Record Office. D/Rh. 1/2.
64. Davies, Clifford. *The Evolution of Industries and Settlements between Merthyr Tydfil and Abergavenny from 1740 to 1840.* University of Wales thesis, 1949.

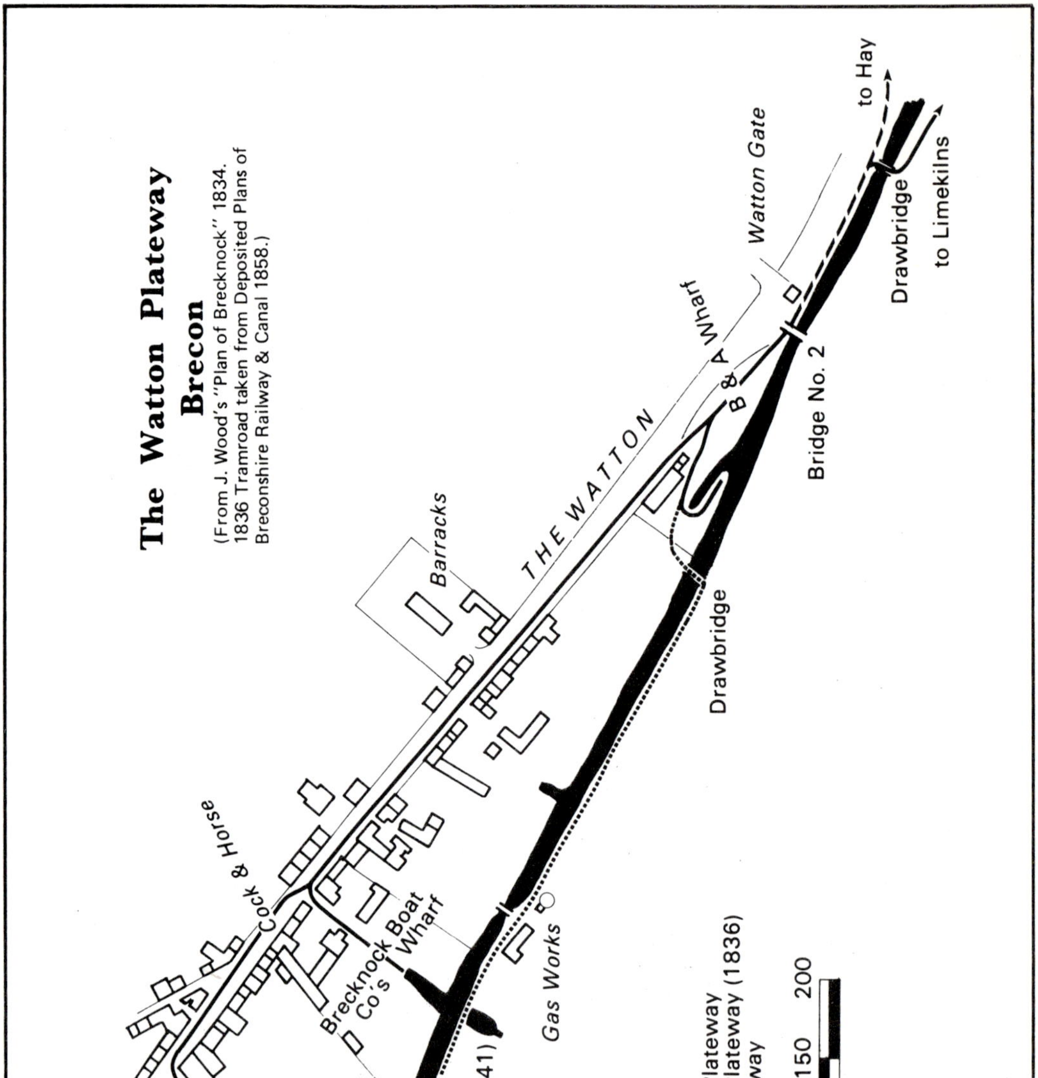

The Watton Plateway Brecon

(From J. Wood's "Plan of Brecknock" 1834. 1836 Tramroad taken from Deposited Plans of Breconshire Railway & Canal 1858.)

11

The Watton Plateway

When the Brecknock & Abergavenny Canal was completed to Brecon in December 1800,[1] the company established wharves at The Watton (SO.0512 2803) and at the canal head (SO.0457 2835). In the previous January, the Brecknock Boat Co. had leased land at *The Cock & Horse* (SO.0475 2820), on the proposed line of the canal, as a site for a wharf,[2] to be ready when the canal should reach the town. Hugh Bold also established a wharf on land he owned on the terminal arm at SO.0459 2836, opposite the B&A town wharf.

When, in 1805, a scheme was mooted to construct a tramroad from Brecon to Hay, amongst the subscribers were two of the partners in the BBCo., Jeffreys Wilkins, the Brecon banker, and John Lloyd,[3] both of whom were shareholders in the B&A. This scheme failed to get off the ground, but in 1810 it was revived by a fresh body of promoters, but still with the same representatives of the BBCo. appearing in the subscription list.[4] The BBCo. was confident that this later attempt would have positive results, but thought its southern terminus would be at Brynich Bridge, in accordance with the plan prepared by William Crosley, then engineer to the B&A. Accordingly, on 2 July 1810, John Peirce, another partner, purchased a plot of land on the canalside at SO.070 276 on their behalf for £900.[5] Had the tramroad been constructed to this plan, the site would have been particularly advantageous, with one frontage on the canal and one on the Brecon to Abergavenny turnpike, with the tramroad running into it. Apart from easing the distribution of its wares, it would have enabled a saving of one and three-quarters of a mile of canal tonnages to have been made on all cargoes to the east and north of Brecon.

In the event, due to opposition from the B&A, the terminus of the tramroad was located at that company's Watton Wharf. The BBCo., and Hugh Bold, and his tenants, Wm. Powell & Co. were therefore left with the choice of using their own wharves, and paying for transhipment from boats to carts and from carts to trams to get their goods onto the Hay Railway, with the risk of breakage of coal, or of paying wharfage to the B&A at Watton.

The Hay Railway was, officially, declared open on 7 May 1816,[6] and when the canal committee met on 19 July, they had before them a propostion from the BBCo. and Hugh Bold, jointly, that a tramroad should be constructed from Watton Wharf to the canal head to serve Bold's wharf, and with a branch into that of the BBCo. They proposed to finance the construction themselves, and on being repaid by the B&A they would lease it from the company at five per cent of the cost as yearly rent. The cost was estimated at not more than £400. The B&A committee concurred and suggested that, in view of the depressed state of trade, and the unemployment in the district it would be of advantage to put the work in hand as soon as possible, in any case it should be done before the ensuing harvest absorbed the surplus labour and forced up wages.[7] The decision was confirmed at the assembly in October, which additionally sanctioned the construction of a short line round the basin at Watton Wharf to connect with the Hay; it is surprising that this was not thought of when the Hay was under construction.

The committee of 27 January 1817 proposed another short tramroad from the Hay, crossing the canal by a draw-bridge at SO.0521 2797, to the B&A-owned kilns at SO.0550 2795, to facilitate the carriage of lime to places served by the Hay. Several of the kilns were let to the BBCo. and at least one to William Powell & Co., and it was considered that the provision of a tramroad connection would prove advantageous when next the letting of the kilns arose. The scheme met with wholehearted approval at the Hay committee meeting on 26 February to the extent of offering to supply the plates necessary[8] and with this endorsement of their plan, the B&A ordered the line to be laid immediately.

The tramroad from Bold's wharf to the Hay was apparently completed by 9 November 1816, as the lease to the BBCo. and Bold (dated 22 March 1817) ran for twenty-one years from that date; the rent of £17 a year was to be shared equally between the two wharf-owners.[9] The line was to be considered 'public', and the lessees were to have the benefit of any tonnages paid by outside parties, including those arising on the part lying in Watton Wharf. This, taken in conjunction with the fact that the B&A had to build their own line round the basin at Watton, would indicate that the Hay Railway terminated either at, or only a few yards to the west of, canal bridge No. 2 at SO.0518 2800, and did not actually serve the wharf.[10]

That the annual rental was £17 shows that the cost must have been £340, which the B&A had to pay. Instead of settling straight away, Bold and the BBCo. were each given promissory notes for £120 payable after twenty-one years with interest.[11] They were, in fact, paid on 25 January 1838 and are stated in the minute book to have been then valued at £170 each.

The B&A realised that the tramroad might be useful to them in making a communication between their two wharves, and in March 1817, they entered an agreement with Bold to have the use of the line through his wharf in order to reach their own, at an annual rental of £6 for twenty-one years. They do not appear to have made much use of this facility, and on 28 August 1820 a sub-committee which had been investigating ways in which canal expenditure might be reduced suggested that, as the tramroad " . . . is not now used . . . we recommend that the rent paid to Mr. Bold be discontinued . . . " The committee, however, decided that the company would be obliged to honour its agreement.

In spite of the enthusiasm of the moment, the tramroad around the basin at Watton Wharf, ordered to be made in October 1816, did not materialise, and at a meeting of 26 December 1820 it was reported that the wharf was too small to contain stock " . . . unless the Tram Road therein be extended to the wharf round to the opposite side of the Basin . . . ", and once again, the Agent was ordered to get the extension made.

Trade was bad in the years between 1820 and 1825, in the aftermath of the Napoleonic wars, and the tramroad was apparently allowed to deteriorate, so much so that the canal assembly received a request on 28 April 1825 from the borough authorities of Brecon, to have it removed altogether. It fell to the lot of the committee to deal with the matter, and at the beginning of June the Agent was instructed to ascertain what action was called for. By the end of the month he had submitted his report, and the committee decided that the position would best be rectified by " . . . putting a string course of good quarry stones, raising it a little above the elevation of the Rails, and filling and packing the same with well-broken Lime Stone . . . ",[12] which would render it safe for the public and preserve the essential communication. It was all very well for the committee to decide on what action to take, but at the next meeting, on 21 July, it was found on examination of the lease to Bold and the BBCo. that the lessees were, in fact, liable for the upkeep, and not the B&A, and the matter was passed to them for settlement. There is no record of their taking any action.

The tramroad was again causing bother in 1833, and the assembly of 25 April received complaints of the state of the line. It was decided that the only way out of trouble would be to remove it to the other side of the canal, and lay it on the tow-path, with drawbridges into the various wharves. Remembering their previous experience, the resolution was worded, "... that the lessees of such Rail Way be allowed to divert it . . . "

There was considerable delay before any action was taken, and John Wood's "Plan of Brecon" dated 1834[13] shows the tramroad running the length of The Watton, as the street from the Watton turnpike gate into the town is still known. It is not difficult to understand the annoyance that the line must have caused in a busy street, the only access to the town from the east. The plan shows that from Watton Wharf it ran on the south of the roadway, nearly to the turning into the BBCo's. wharf at *The Cock & Horse,* but before the turnout into their premises it crossed to the north of the street and ran on that side to the turning into Bold's wharf, making two places where it cut across the road traffic.

Possibly because the lease was nearly up, the B&A committee decided on 26 October 1835, to take matters into their own hands, and indeed, it is difficult to see how they could have enforced the change to a new site on the lessees. The Agent was instructed to proceed "... commencing at the northern extremity of the Company's Upper Wharf... and terminating at the Head of the Basin in the Company's Lower Wharf... " It was specified that the cost should not be more than £300 plus whatever sum could be obtained for the old plates. The estimate and plans were prepared by Robert Smith, the mineral engineer to the Clydach Ironworks, and his estimate shows a considerable excess over the limit stipulated.[14] It is divided into two sections, one showing the cost of the alterations actually within the company's wharf quotes £61 15s. for preparing the track and the draw-bridge over the canal, plus the cost of plates, £27; the other for the 'public' road amounts to £132 2s. 3d., with plates as an additional £186 15s. a total of £407 12s. 3d. The latter estimate only allows for one draw-bridge over the canal, presumably that leading into the BBCo's. wharf. No date has been found for the completion of the alterations.

In 1841 the new line was in need of repair and £50 was voted for this purpose at the meeting of 21 October, and a month later the weighing machine was reported to be out of order, and the Agent was instructed to obtain estimates for its repair. As the Agent, James Peirce, undertook the contract for repairs to the canal and railroads in July 1841, it is assumed that he was responsible for the work, although the minute book makes no reference to his undertaking repairs to a "tramroad".

The plateway can hardly have been profitable to the B&A, as the receipts from 1851-59 averaged only £6 a half-year, with a maximum in the first half of 1853 of £16 and a minimum of £4 in the second half of 1858. No doubt, in its early days, it saved the BBCo. more than the £8 10s. a year that they were paying, but probably, when it reverted to the B&A, and they had to pay tonnages for its use, it proved more economical to pay wharfage and land their cargoes for the Hay Railway at Watton Wharf, using the tramroad made from the canal side round the basin.

The deposited plans of the Breconshire Railway & Canal Co. of 1858 for the parish of St. John the Evangelist, Brecon,[15] show the tramroad on the canal towpath, but do not show it continuing beyond the BBCo's. wharf, and whilst the draw-bridge to the latter is shown, the tramroad across it is not, and it would appear that the line was maintained principally to serve the gasworks at

SO.0548 2798. The plan does show the branch from the Hay Railway to the kilns crossing the canal at SO.0548 2798.

The line to the kilns was still in existence in 1861, and in October Richard Pruett, who had contracted to repair the B&A property in 1851, with a renewal in 1856, was granted an additional £5 5s. a year for keeping in repair "... that portion of the Hay Railway belonging to this Company at Brecon ...",[16] but in January 1862 it was decided to dispense with the line, and orders were given for the plates to be lifted and stored, but a year later, it was ordered that they should be sold provided that £3 per ton could be obtained for them. Presumably, with the opening of the Brecon & Merthyr Railway imminent, it was considered that the era of the plateway was over.

No further references have been found to the B&A tramroads in Brecon, and it can only be assumed that the Watton Plateway was left to decay in the course of the years.

Appendix

Probably in the first year of its working the Watton Plateway was operated under the by-laws applicable to the B&A's Clydach Rail Road, but the general assembly of 29 April 1819 passed a set of by-laws specifically applicable to the Brecon tramroads, as under:

1. Ordered that no Waggon or other Carriage be permitted or allowed to pass along the Rail Way of this Company at Brecknock with a load exceeding Two Tons weight, nor unless the Wheels and Axles thereof be fitted to the Gauge of the Rails and the Waggons constructed according to the directions of the Committee.
2. That no Driver of a Waggon or other Carriage be permitted to draw the same off the Rails unless such Waggon or other Carriage is disabled by an Accident from proceeding to its destination, in which case, the same shall be forthwith unloaded and moved off the Rails.
3. That no Driver shall be permitted or allowed to Ride upon his Waggon or to draw it on the said Rail Way faster than a walking pace.
4. That no Driver of a Waggon or other Carriage shall unnecessarily halt his Horse or Horses so as to obstruct the Passage of the said Rail Way.
5. That no Person be allowed to impede or obstruct the Passage of the said Rail Way, or leave their Waggons or Goods to be there, or to throw any Rubbish or Stones upon the same.
6. That if any Waggon or other Carriage break down upon the said Rail Way the same shall be immediately removed off the Rails.
7. That all empty Waggons shall draw back to the next Parting and give way to Loaded ones passing along the said Rail Way.
8. Ordered that any Person or Persons offending against any of the above Orders shall for every Offence Forfeit and Pay a sum not exceeding Forty Shillings.
9. The foregoing are declared to be Bye-Laws.

REFERENCES:

1. Hadfield, Charles. *Canals of South Wales & the Border*. D&C. 1960.
2. GRO. D.591/11.61.
3. *Hereford Journal*, 29 May 1805.
4. ibid. 25 July 1810.
5. Brecknock Museum, Brecon. Maybery Collection, Vol. 1A. fo. 17.
6. Herefordshire & Worcestershire R.O., Hereford. Hay Railway minute book quotes 7 May 1816; the *Hereford Journal*, 22 May 1816 implies that the opening date was 16 May 1816.
7. PRO. B&A committee minutes, 19 July 1816.
8. Hay Railway minutes, loc. cit. 26 February 1817.
9. NLOW. Maybery 981.
10. ibid. and see GRO. D.591/11.61.
11. PRO. B&A committee minutes, 27 January 1817.
12. ibid. 27 June 1825.
13. NLOW. Map Room. Published by John Wood, Caernarvon.
14. NLOW. Maybery 1139.
15. NLOW. Map Room.
16. PRO. B&A committee minutes, 17 October 1861.

Acknowledgements

In the course of research, particularly when it is spread over a number of years, one meets and talks with many people who have information that they are willing to impart. It is unfortunate, that whilst one may know their names at the time, in the course of years they tend to become increasingly anonymous. To those in this category I tender my sincere apologies for my forgetfulness, and I would assure them that I am, nonetheless, exceedingly grateful for their co-operation.

The archivists and their staffs of Gwent and Glamorgan Record Offices have always been generous with their help, as have the Curator and staff of the Brecknock Museum, Brecon, and I am deeply grateful to them.

My thanks must also go to the staffs of the National Library of Wales, Aberystwyth and the British Transport Historical Records branch of the Public Record Office.

Thanks must be given to His Grace, the Duke of Beaufort for permission to examine the Badminton Papers held at Aberystwyth, which, although none has been quoted directly, have provided clues to be followed up in other places.

The advance of money by the Railway & Canal Historical Society has been invaluable in helping with the expenses incurred in travelling and hotel accommodation whilst doing research.

Without the prodding I have received from Messrs. P. A. Stevens and H. W. Paar, and the late D. Morgan Rees, Keeper of the Industry Department of the National Museum of Wales, Cardiff, I doubt if I should have had sufficient self-confidence to undertake writing the foregoing pages, and I can only thank them for their faith in me.

Mr. John van Laun, of Longtown, Herefordshire, must receive special mention, as not only has he put his keen eye at my disposal when on fieldwork, but he has made many helpful suggestions for improvement of my manuscript, and has undertaken the self-appointed task of checking all my map references.

Lastly, my gratitude to my wife, who, completely uninterested in the subject matter, has, perforce had to listen to the clack of my typewriter for periods, which over the years, must have seemed interminable.

10 Orchard Crescent,
Dinas Powis, South Glam.
July 1979.

Gordon Rattenbury.

Index